Make: Analog Synthesizers

Ray Wilson

SEBASTOPOL, CA

Make: Analog Synthesizers

by Ray Wilson

Printed in the United States of America.

Published by Maker Media, Inc., 1005 Gravenstein Highway North, Sebastopol, CA 95472.

Maker Media books may be purchased for educational, business, or sales promotional use. Online editions are also available for most titles (*http://safaribooksonline.com*). For more information, contact O'Reilly Media's corporate/institutional sales department: 800-998-9938 or *corporate@oreilly.com*.

Editor: Shawn Wallace	**Indexer:** Jill Edwards
Production Editor: Kristen Borg	**Cover Designer:** Karen Montgomery
Copyeditor: Charles Roumeliotis	**Interior Designer:** David Futato
Proofreader: Amanda Kersey	**Illustrator:** Robert Romano

May 2013:   First Edition

Revision History for the First Edition:

2013-04-29:   First release

2013-08-16:   Second release

2014-08-29:   Third release

See *http://oreilly.com/catalog/errata.csp?isbn=9781449345228* for release details.

ISBN: 978-1-449-34522-8

[LSI]

For Debra Lee, whose love and kindness sustain me; Jonathan and Arielle, the lights of my life; and my parents, Charles and Elia Wilson, who always encouraged me to be inventive.

Table of Contents

Preface

I hope this book becomes tattered from use and is kept close at hand on your desk or work bench. We're going to cover a lot of information about analog synthesizers and how to get started in the fun and engaging hobby of *synth-DIY*. You'll need some electronics knowledge to fully benefit, but I think you'll find what's here interesting and informative regardless of your experience. Be forewarned that this is one of those hobbies that can become consuming, with ideas building on other ideas and inspiration coming at all hours of the day and night. You may find yourself soldering in your skivvies from time to time because you come up with an idea that simply can't wait for tomorrow to try out.

How I Got Started

I've had a consuming interest in analog synthesizers since the first time I heard *Switched-On Bach* back in the spring of 1968. Back then, the only thing that separated me from one of those analog beauties was cash, and lots of it. Those early synthesizers were way out of my financial reach, as in thousands of dollars out of my reach. However, I had an excellent opportunity to at least check them out and play them. My window into the world of analog synthesizers was provided by a music store in McKees Rocks, Pennsylvania, that has been there since 1965.

I haunted the keyboard room at *Hollowood Music and Sound* just about every day of my adolescence, and I can still remember being able to try out any of the amazing units on display. Looking back, I really appreciate how cool Mr. Hollowood was for allowing a skinny kid with long, curly hair and a penchant for trying to imitate Keith Emerson (I said trying…) to jam on any synth in the store while he was well aware that I couldn't possibly afford any of them (Figure P-1). The person who ran the keyboard room was an amazing keyboard player named John Bartel. John was one of my heroes because not only was he an outstanding keyboardist, but he actually knew how to get an awesome sound out of any synth in the place, including the ARP 2600.

Figure P-1. *Would you let this character try out your ARP 2600?*

In those days I didn't know an ADSR from the business end of my alimentary canal. Eventually, after I started to work at U.S. Steel (as in an actual steel mill), I had some money in my pocket, and you can probably guess where I spent it. I went through a litany of analog synthesizers, buying, selling, and trading mainly Korgs, Mini-Moogs, and a variety of the patchable semimodular Roland units. I'm ashamed to say that often I didn't fully explore the nuances of a synth before trading it off for a newer, shinier one.

Analog synth greed got a hold on me, oh lordy it got a hold on me. Playing and recording with them on my Tascam 4-track with 10-inch reels was a blast, but I wanted something more. I wanted to be able to make an electronic device that would allow me to interact with the synthesizer somehow. I had grandiose visions of someday making my own synthesizer. Little did I know that one day I *would* be building my own modular analog synthesizers that could do anything any synthesizer in that keyboard room could do (and more). And *you can too*.

My Electronic Roots

When my father, who was a remarkable mechanical engineer, used his influence to get me a job at the U.S. Steel Wheel & Axle Works in McKees Rocks, Pennsylvania, as a laborer in the "labor gang," I couldn't have guessed what was in store. To this day it surprises me that the mill would eventually be my ticket to receiving electronics training and in time (OK a *lot* of time) lead to me starting my own small electronics business, *Music From Outer Space LLC*, which I run to this day.

After a year or so of shoveling, sweeping, and jackhammering—and way more than likely at the behest of my father—I got into an electronic repairman apprenticeship program. This meant that I was freed from the labor gang! Not that jackhammering scale (the drossy crust that falls from steel billets as they are being heat-treated) out of furnaces that had been heat-treating train wheels the day before wasn't fun, but the apprenticeship was a godsend (well actually a *dadsend*). The apprenticeship meant that I got to go to the Homestead Works in Homestead, Pennsylvania (where they used to produce those giant buckets full of molten steel) and get paid more than I did as a laborer to learn electronics!

I had found my passion. I supplemented my classroom training with breadboard experimentation, a Heathkit microprocessor trainer kit, and reading, reading, and then… more reading. I did the practice math problems and experiments in the books, and I'd encourage you to do the same if you're just getting started. Through it all I learned enough to spend the next 15 years in the medical electronics industry, finally landing in the biomedical engineering department of Siemens Pacesetter. Interestingly, I was able to do this without an engineering degree, but I think those days are over, so get in college and stay there until you're armed and dangerous with a degree in something useful.

Equipped with enough electronics training to be dangerous, a friend and I started a small electronics kit sideline called *Waveform Processing*. We were located in Crafton, Pennsylvania, and were in business from the late 1970s into the early 1980s. I designed some simple sound-making boxes (Figure P-2 shows the construction book and PC board that came with the Waveform Processing Mini-Synth), etched boards, kitted parts, and advertised in the much revered and renowned *Radio Electronics* magazine.

Figure P-2. *The Waveform Processing Mini-Synth*

To this day, I still cringe when I think of what we used to pay for a 15-word classified ad buried among hundreds of others in the back of that magazine: $500. Needless to say, the brand "Waveform Processing" is not rolling off of anybody's tongue today.

My interest in electronics didn't wane by any means; I kept on researching and breadboarding and developed some very cool mono- and polyphonic synths. I remember the excitement of discovering the Curtis Electromusic Specialties chips designed by Doug R. Curtis. They were simply the coolest chips ever. You could easily build voltage-controlled filters and voltage-controlled oscillators that tracked one volt per octave perfectly and were completely temperature compensated. I was in heaven. Eventually, however, family and job responsibilities caused me to pay less and less attention to synth-DIY.

I was getting my electronics design fix at Intec Systems in Blawnox, Pennsylvania, and eventually Siemens Pacesetter in Sylmar, California, so my focus shifted to designing electronic test equipment and writing software. In 1994, an earthquake of magnitude 6.7 chased us out of California and on to Aurora, Colorado. I can still remember throwing my first attempt at a modular into a huge dumpster as we hurriedly cleaned house and prepared to move. *(Man I wish I still had that thing.)*

After my work focus shifted entirely to software development, my *need* to work with electronics brought me full circle and back to synth-DIY with a vengeance. I started a website called Ray-Land (perhaps I'm a tad bit narcissistic) and began to publish the circuits, PC layouts, and circuit descriptions I was coming up with. I developed a simple, battery-powered mini-synthesizer called the *Sound Lab Mini-Synth* and published that as well, and it started to attract attention (Figure P-3 shows the very first Sound Lab Mini-Synth prototype).

Figure P-3. *The* Music From Outer Space *Sound Lab Mini-Synth prototype*

Music From Outer Space

A fellow DIYer suggested I start selling the PC board for the Sound Lab Mini-Synth, and before you knew it, my current business, *Music From Outer Space LLC*, was born. I had the privilege of writing a 10-page article about the Sound Lab Mini-Synth for the March 2006 edition of the magazine *NUTS AND VOLTS, Everything For Electronics*. They even paid me half-decently to write it, to boot. I get a kick out of it to this day.

Figure P-4. *The Music From Outer Space website*

The name *Music From Outer Space* (also known as MFOS) was inspired by the fact that space program–related research was instrumental to the successful development of transistor and integrated circuit (IC) technology. Eventually, the technology trickle-down brought highly reliable, reasonably priced electronic components, transistors, and ICs to the masses. Today, MFOS (*http://www.musicfromouterspace.com*) is a popular website visited by synth-DIYers the world over (Figure P-4 is a screenshot from the website).

It's a great feeling to receive email from visitors expressing appreciation for the information and resources the site provides. I never forget how I learned from and was inspired by the work and invention of the synthesizer pioneers, many of whom are sadly gone now. The site's *Acknowledgments* page pays homage to my synthesizer heroes: the analog synthesizer inventors and electronics book authors who have inspired and taught so many of us. I'm convinced that synth-DIYers are some of the most creative and clever people around today, and I'm happy to say that many of the emails I receive from people asking questions or kindly thanking me are from some of the world's most prestigious learning institutions.

What You Should Know

If you have no electronics knowledge or training but still have a burning desire to get into analog synth-DIY, you have a bit of work ahead. You'll need to understand basic electronic principles such as:

- Reading a schematic

- Ohm's law
- How passive components work: resistors, capacitors, coils, transformers, switches, and relays
- How active components work: diodes, transistors, integrated circuits, and LEDs I heartily recommend the book *Make: Electronics* by Charles Platt (O'Reilly/MAKE). Don't think you have to completely understand the quantum mechanical level of operation before you can even get started, or you may *never* get started. The quantum community is still duking it all out.

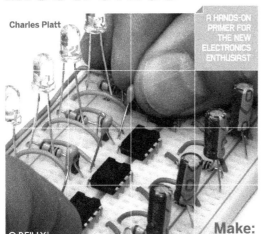

Burn things out, mess things up—that's how you learn.

Make:
Electronics

Learning by Discovery

Charles Platt

A HANDS-ON PRIMER FOR THE NEW ELECTRONICS ENTHUSIAST

O'REILLY°   Make:

Figure P-5. Make: Electronics is an excellent kickstarter for your basic electronics knowledge

Purchase one of those 75-in-one electronic experimenter kits that have the components connected to springs or clips that facilitate building up and tearing down circuits quickly. The kits normally come with a book that takes the user through scores of experiments that often build on one another and clearly present the principles of operation for each circuit presented. I suggest

you discipline yourself to simply step through *every* experiment and maybe even do each one *twice* for good measure! Yes, the kits look a little cheesy in their colorful cardboard boxes, but you can bet that Bobby Moog himself went through a number of these in his youth, and believe me, if you apply yourself, you *will* learn quite a bit.

The material presented in this book assumes a basic level of electronics knowledge and experience, including schematic reading, recognizing electronic components and how to read their values, soldering, and constructing electronic projects, including PC board population and front panel wiring. And no matter if you've been building for years or if you've just started, you'll need to know some troubleshooting techniques.

Once you've got the electronic basics down, you're ready to start finding synth-DIY projects to experiment with. The Music From Outer Space website (*http://www.musicfromouterspace.com*) presents scores of fully documented synth-DIY projects for both newbies and advanced hobbyists. I think you'll find the Noise Toaster lo-fi noise box (see Chapter 4) to be a great starting project if you're a synth-DIY newbie—and just one of the coolest portable toys ever if you already have DIY skills. It makes a really wide variety of sounds, and it's a solid analog synthesizer through and through.

What's Next?

Chapter 1 gives some background and shows a few examples of what kinds of synthesizers are possible to build once you have some synth-DIY experience under your belt. Chapter 2 shows you what tools, test equipment, and electronic components you'll need in your work area or lab to succeed in your synth-DIY work. Chapter 2 also covers how to improve your soldering skills and presents some sage troubleshooting advice. In Chapter 3 I'll discuss the purpose and operation of the various analog synthesizer modules, what they do, and how they interconnect and interact. Chapter 4 describes the construction of the Noise

Toaster analog noise box project. Chapter 5 presents an introduction to op amps that leads into an explanation of the circuitry behind the Noise Toaster in Chapter 6. Chapter 7 brings it all together with some useful information about how to record your next platinum synth music CD using your computer as a multitrack recording studio. The three appendixes dive deeper into several common electronic circuits found in analog synthesizer modules, how they function, and how you can use them in your own projects.

We've got a lot of material to cover, so let's get started.

Conventions Used in This Book

The following typographical conventions are used in this book:

Italic

> Indicates new terms, URLs, email addresses, filenames, and file extensions.

> *This box signifies a tip, suggestion, general note, or warning.*

Using Examples

This book is here to help you get your job done. In general, if this book includes code examples, you may use the code in this book in your programs and documentation. You do not need to contact us for permission unless you're reproducing a significant portion of the code. For example, writing a program that uses several chunks of code from this book does not require permission. Selling or distributing a CD-ROM of examples from MAKE books does require permission. Answering a question by citing this book and quoting examples does not require permission. Incorporating a significant amount of examples from this book into your product's documentation does require permission.

We appreciate, but do not require, attribution. An attribution usually includes the title, author, publisher, and ISBN. For example: "*Make: Analog Synthesizers* by Ray Wilson (MAKE). Copyright 2013 Ray Wilson, 978-1-449-34522-8."

If you feel your use of code examples falls outside fair use or the permission given above, feel free to contact us at *permissions@oreilly.com*.

Safari® Books Online

Safari Books Online is an on-demand digital library that delivers expert content in both book and video form from the world's leading authors in technology and business.

Technology professionals, software developers, web designers, and business and creative professionals use Safari Books Online as their primary resource for research, problem solving, learning, and certification training.

Safari Books Online offers a range of plans and pricing for enterprise, government, education, and individuals.

Members have access to thousands of books, training videos, and prepublication manuscripts in one fully searchable database from publishers like O'Reilly Media, Prentice Hall Professional, Addison-Wesley Professional, Microsoft Press, Sams, Que, Peachpit Press, Focal Press, Cisco Press, John Wiley & Sons, Syngress, Morgan Kaufmann, IBM Redbooks, Packt, Adobe Press, FT Press, Apress, Manning, New Riders, McGraw-Hill, Jones & Bartlett, Course Technology, and hundreds more. For more information about Safari Books Online, please visit us online.

How to Contact Us

Please address comments and questions concerning this book to the publisher:

> Maker Media, Inc.
> 1005 Gravenstein Highway North
> Sebastopol, CA 95472

800-998-9938 (in the United States or Canada)
707-829-0515 (international or local)
707-829-0104 (fax)

We have a web page for this book, where we list errata, examples, and any additional information. You can access this page at *http://oreil.ly/make-analog-synthesizers*.

To comment or ask technical questions about this book, send email to *bookquestions@oreilly.com*.

Maker Media is devoted entirely to the growing community of resourceful people who believe that if you can imagine it, you can make it. Maker Media encourages the Do-It-Yourself mentality by providing creative inspiration and instruction.

For more information about our publications, events, and products, see our website at *http://makermedia.com*.

Find us on Facebook: *https://www.facebook.com/makemagazine*

Follow us on Twitter: *https://twitter.com/make*

Watch us on YouTube: *http://www.youtube.com/makemagazine*

What Is Synth-DIY?

1

In the mid-1960s an electronics engineer with a Ph.D. from Cornell University in engineering physics was starting a musical revolution, and he didn't even know it. That engineer was Robert Moog. In a small factory in bucolic Trumansburg, New York, he was developing a series of unique electronic analog sound synthesizers that would take the musical world by storm.

At the same time, the muse of invention was striking a West Coast engineer named Don Buchla. Buchla was developing his own unique series of electronic analog sound synthesizers. Both Moog and Buchla's remarkable inventions were *voltage-controlled*; they both used a varying voltage to control the sound-shaping functions of the synthesizers' modules.

It wasn't long before analog synthesizer sounds and music were being heard everywhere all the time. They made their way into music studios, records, commercials, jingles, company sound logos, movies, every genre of music, video games: everything. Analog synthesizer sounds and music had become so widespread that scores of companies saw the potential market and dove headlong into the analog synthesizer business. Some popular manufacturers were Moog, ARP, Oberheim, Yamaha, Korg, Fairlight, E-mu Systems, Roland, and many others.

Although extremely popular for many years, the analog synthesizer's sales numbers eventually began to dip with the advent of less expensive, easier-to-manufacture, *and pretty darn snazzy* digital synthesizers. Instead of creating sounds using analog sound-producing circuitry, digital synthesizers use microprocessors to do the heavy lifting. Binary representations of various sound and instrument waveforms are stored in read-only memory (ROM). The microprocessor coordinates the synthesizer's real-time user in-terface activity, scans the polyphonic keyboard, and finally processes and streams the digital audio data to the synthesizer's output.

This was one of the digital synth's weaknesses. Instead of the completely continuous, organic, full sound of the analog synthesizer, the digital units often included digital conversion artifacts and lacked sonic character. Figure 1-1 highlights the difference between a waveform generated digitally and a continuous waveform generated by analog circuitry. The early digital units borrowed from analog synthesis for filtering and sound envelope shaping for a time, but the lower cost, manufacturing advantages, and technical advances of the digital units soon ended the analog synthesizer's dominant reign of many years.

Sadly, history also shows that while it takes an electronics genius to invent amazingly cool synthesizers, it also takes a great deal of business sense to keep a company in the black. Just about all of the major players in the synthesizer industry had an electronic genius or two on board, but only a few remembered to hire a chief financial officer (CFO). Many of the major players in the synthesizer business are now only a memory, a distant electronic wail in the breeze.

However, analog synthesizers were just too cool to stay gone forever. There has always been a small analog synth priesthood keeping the faith alive, but today, people all over the planet are rediscovering the warm, inimitable analog synthesizer sound. They're finding that turning *real* knobs, flicking *real* switches, plugging in *real* patch cords to produce *real analog* sounds is way more interesting and engaging than scrolling through an LED parameter display or clicking virtual switches with a mouse. Today many musicians and hobbyists with an interest in electronics are discovering that with the right schematics, tools, printed circuit boards (PCBs) and ingenuity, making their own analog synthesizer is well within reach. The book you're holding in your hands right now has the information you need to put you squarely onto the road to *doing it yourself* when it comes to building your very own analog synthesizer.

A Bit of Analog Synthesizer Etymology

Before we move on, let's take a moment to consider what the term *analog synthesizer* really means. The word *analog* puts me in mind of *organic*, *living*, *real*, and *continuously variable*. A good definition of the word *analog* would be "of, relating to, or being a mechanism in which data is represented by continuously variable physical quantities." The key point we're interested in as far as describing an analog synthesizer is "continuously variable." An analog synthesizer creates sound by means of electronic circuit elements: resistors, capacitors, transistors, and integrated circuits. If you observed the output of an analog synthesizer with an oscilloscope, you would see that the waveform voltage is continuous, as opposed to the characteristic stepped appearance of a digital signal source (Figure 1-1). This is one of the characteristics that give analog synthesizers their warm, fat, and much-sought-after sound.

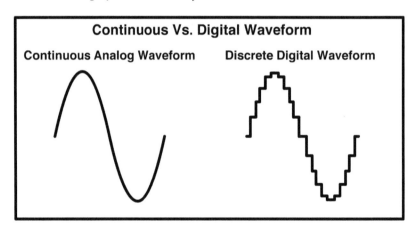

Figure 1-1. *Continuous waveform versus discretely stepped digital waveform*

Additionally, much of the circuitry found in an analog synthesizer has roots in the early analog computers. These were very early computers built from operational amplifiers and passive electronic components (resistors, capacitors, coils, etc.) that performed mathematical operations such as logarithms, multiplication, division, addition, subtraction, and even root finding. Analog computers were used by scientists and engineers to solve all manner of mathematical problems before modern digital computers were developed. Variables were entered using potentiometers (informally known as *pots*) to set voltages, which the analog computer would process in real time. Interestingly, the circuit used in an analog synthesizer to regulate the response of the voltage-controlled oscillator (VCO) to control voltage is a linear voltage to exponential current converter that might well have been found in an electronic analog computer.

Now we turn to the term *synthesizer*. To synthesize is to take existing elements and combine them to produce something new. Anyone who has heard the incredible range and variety of sounds that come from an analog synthesizer would have to agree that its combination of unique electronic circuitry produces incredible sounds not heard before on this planet. Analog synthesizers are also perfect for producing convincing reproductions of just about any acoustic or electronic instrument.

How do I define analog synthesizer-DIY, or synth-DIY, as I call it? It's people making their own electronic sound boxes, noise makers, and synthesizers; and modifying Speak & Spells and other vintage electronic sound toys to get weird and unusual sounds from them. It's conventions where from a few to as many as hundreds of people who share the synth-DIY passion get together to compare notes and see and hear one another's latest projects. It's learning about electronics and components, reading schematics, soldering PC boards, and wiring front panels. It's learning where to find the best prices and selection for electronic components. It's learning what tools and equipment you'll need to succeed and how to troubleshoot the projects you make. It's being creative, including learning how to set up a recording studio right on your computer so you can make multi-track recordings of your sound creations. And finally, its actually making your own analog synthesizer, from something as small as a lo-fi noise box to as huge as a modular monster that would have cost tens of thousands in the analog heydays.

What Can I Build?

What you can successfully build depends on the level of electronics knowledge you possess, the effort you put forth to learn and become proficient, and the passion you have to build your own analog synthesizer. Please rest assured that even

if you are just getting started, the analog modular *sky* is the limit—if you really want to get there. I always suggest that newbies cut their teeth on solderless breadboard experimentation or one of those electronic experimenter kits. After you've gone through the experiments and know a bit more about resistors, capacitors, transistors, and ICs, step up to a kit with a PCB that requires soldering—because wielding a soldering iron is a fundamental skill you just can't do without. I'll go over the tools you'll need and how to set up your workbench in Chapter 2. But now, let's take a look into your possible DIY future and feast our eyes on some photos of analog synthesizer DIY projects people have contributed over the years. Many graphic artists are drawn to synth-DIY and express themselves not only audibly but visually as well.

Figure 1-2 shows the popular MFOS *Alien Screamer* built by DIYer Bernard Magnaval, of France. DIYers build their own cases and make their own faceplates. I love the creativity shown in this build: the bright colors and cool-looking knobs. The Alien Screamer has a VCO and a low-frequency oscillator (LFO) that can be used to both modulate and sync the VCO. The LFO provides a variety of waveforms with which to modulate the VCO's frequency. For its size and simplicity, this little guy makes some totally cool sounds.

Figure 1-2. *Music From Outer Space Alien Screamer*

This DIYer etched the circuit board for this eight-step sequencer in Figure 1-3 from a layout published on the MFOS website and then built up the entire project. Sequencers are used for repeating patterns of notes and creating arpeggiation effects.

Figure 1-3. *Music From Outer Space eight-step sequencer*

The Sound Lab Mini-Synth (Figure 1-4) is a favorite among DIYers, with its two VCOs and warm-state variable voltage-controller filter (VCF), as well as a normalized switching scheme that permits a *ton* of sonic variety. This is the project that put MFOS on the synth-DIY map and has been built by hundreds of people around the world.

Figure 1-4. *Music From Outer Space Sound Lab Mini-Synth*

Some people go to great lengths to express themselves through synth-DIY. Figures 1-5 through 1-7 show some of the more unusual examples. If you plan to gut your Stradivarius for a synth case, I suggest you think it over calmly for a bit, but in the end, the decision is yours. Using parts of a human body is also up to you, but as you can see, it has been done. And finally, some folks like to wear their synthesizer and march through their neighborhood. Doing so is entirely up to you and perhaps your homeowners association.

Figure 1-5. *A Stradivarius made into a Weird Sound Generator*

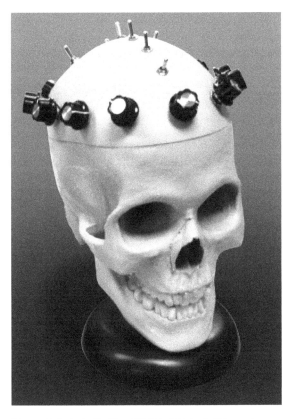

Figure 1-6. *A skull with a WSG built into it; this controversial implant technique is still in clinical trials*

Figure 1-7. *A wearable synth designed for a synth marching band (I would start going to high school football games if the band included WSGs like this one)*

How Far Can I Go?

As I've stated above, and depending on your effort, the sky really is the limit. While not expensive, the practice does require a bit of expendable income. But then, so does just about any hobby. When you start to get serious about analog synthesizer projects, you will be surprised at the cool things you can build. Let's look at some more advanced projects built by some serious DIYers.

The Sound Lab ULTIMATE is a challenging MFOS project, but the attention to detail and effort expended are rewarded with a fully featured analog synthesizer. This project goes farther than the Sound Lab Mini-Synth and adds a third oscillator, a sample and hold module, a second LFO, and several attenuators used for general signal attenuation and control. The normalized switching scheme permits wide sound variability, and patch points permit synthesists to route signals in unique ways. The project in Figure 1-8 (which includes the keyboard) was built by Magnaval.

Figure 1-8. *Music From Outer Space Sound Lab ULTIMATE*

You can create a monster modular synthesizer that has as many modules as you care to build, like the one in Figure 1-9, built by DIYer Michael Thurm of Germany. This synthesizer has capabilities comparable to units costing tens of thousands of dollars back in the analog synth's heyday. The sound possibilities of something like this are virtually limitless.

So now you know where this road can take you. You can dip your toe into the water and build a simple sound generation box, or you can dive into the deep end and build a giant modular analog synthesizer or, of course, anything in between. I can say from experience that synth-DIY is one of the coolest hobbies going. Not only does it provide interesting and engaging building experiences, but the resulting noise boxes and/or musical instruments you make will also provide endless hours of imaginative and creative fun.

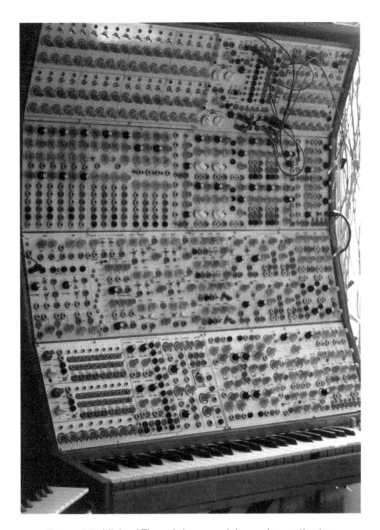

Figure 1-9. *Michael Thurm's huge modular analog synthesizer*

Tooling Up for Building Analog Synths

2

This chapter describes all the electronic instruments you'll need on your workbench to succeed with your synth-DIY projects.

I attended NASA soldering certification classes back in my days at Intec Systems and learned to solder and desolder through-hole and surface mount components under a stereo microscope. I'll share some insights that will help you improve your soldering and, just as important, your desoldering technique.

I'll also share some of the experience I've gained in getting the right price and making the right selections for electronic components. I'll conclude this chapter with some troubleshooting techniques and tips that may save you time once you're done building and have run into problems.

Instrumenting Your Workbench

Trying to work on an electronic project without the proper test instruments is like trying to work on a car without a set of wrenches. It's tough to get the lug nuts off with your fingers. Without a multimeter, you won't be able to read voltages, currents, or resistances. Without an oscilloscope, you won't be able to verify whether a waveform is properly trimmed, a frequency is what it should be, or whether two or more logic signals are coordinated in time.

In some cases, you can get by with a subset if you choose the right tools. For example, a frequency counter is definitely nice to have, but I suggest you save some money by buying a multimeter with frequency counting capability built right in. I bought a decent frequency counter but soon discovered that the frequency counting function in my 50,000-count multimeter captures and displays the frequency using a faster frequency determining algorithm.

As I go over each piece of test equipment, I'm not going to recommend a particular brand or mod-el; instead I'll give you all the features each piece of test equipment should have before you buy it.

The Oscilloscope

There are two types of oscilloscope: analog and digital (Figure 2-1). Both are very useful, but digital scopes have a lot of features that put them ahead of their analog counterparts. *Analog* oscilloscopes display a continuous waveform on their cathode-ray tubes (CRTs), whereas *digital* scopes display their sampled and digitized waveforms on what is essentially a small color computer display. We'll go over the vital specifications for both types and then explore some of the useful functions that only the digital units provide.

Figure 2-1. *An analog (left) and digital (right) oscilloscope*

Frequency Response (or Bandwidth)

All oscilloscopes have a high limit to the frequency they can accurately display without distortion. You'll see scopes advertised as 20 MHz, 40 MHz, 100 MHz, or higher. The higher a scope's bandwidth, the more expensive it is, since they use specialized analog components to attain the higher frequency response. When you try to look at a waveform whose frequency is beyond a scope's bandwidth, it will appear attenuated and distorted such that you just can't trust what you're looking at.

On the low end of the frequency bandwidth, your scope should have DC measurement capability. This is the only way you will be able to look at really slowly changing waveforms or DC voltage levels. A faster scope's Time/Div selector will have smaller time per division settings at the upper end of its range than a slower (lower bandwidth) one. In DC-capable scopes, there is a selector switch for each input channel to set the input mode to AC or DC. In AC input mode, the DC component of the input signal is blocked, allowing you to observe just the AC component of the input signal. This is helpful when trying to observe a low-level signal riding on a relatively high DC level.

The good news is that for analog synth work, you can get away with a relatively low bandwidth analog scope. As a matter of fact, 20 MHz will do just fine. However, if you want to be able to look closely at the edges of digital signals, which do come up even in analog synth work, you may want to go with a 40 MHz, 50 MHz, or even a 100 MHz bandwidth scope. And don't forget to ensure that the scope has DC measurement capability.

The bottom line is this: buy the highest bandwidth DC-capable scope you can afford.

Input Channels

I cut my teeth on a single channel EICO brand oscilloscope that didn't even feature DC measurement capability. Going from no scope to any scope was a great thing. My next step up was a single-channel Hewlett Packard I got for $20 from a friend that at least had DC measurement capability (the scope, not the friend). Single channel will work for you if that's all you can lay your hands on. However, as soon as opportunity knocks, go to a multichannel scope. There are many, many times when you need to see the time difference, voltage difference, and shape difference of two waveforms, and a two-channel scope is just the ticket. Scopes go up to four channels, but those tend to get pretty darn pricey.

I recommend at least two channels for serious analog synth work, but if you happen to own a diamond mine, go for four.

Sensitivity

The vertical sensitivity of an oscilloscope is very important. The better the scope is, the *lower* the sensitivity will be able to go. Again, it takes specialized circuitry to allow a scope to display parts of 1mV per division with any kind of accuracy. That's considered the *noise level* to a lot of people. A decent scope is going to start at about 2mV to 5mV per division but may include a multiplier function that adds a bit more gain and can get you down to a 1mV or better minimum level. The scope's Volts per Division knob is what you use to select the sensitivity, which ranges from about 5mV to 5V, 10V, or higher.

The oscilloscope's probe is an important part of the input circuit (Figure 2-2 shows a typical oscilloscope probe). Scope inputs generally have an impedance of one megaohm. The X1 ("times 1") probe mode is the *what you see is what you're measuring* mode. In addition, most scope probes have a switch that allows them to be set to X10 ("times 10") attenuation. When in X10 attenuation mode, you need to remember to multiply the scope's Volts per Division setting by 10. This is because the probe puts a resistor in series with the scope's input, which causes the voltage measurements to be attenuated by a factor of 10.

This is useful for measuring higher amplitude signals than the scope's Volts per Division setting permits, but just as usefully in this mode, the probe's impedance goes up by a factor of 10 (to 10 megaohms). The higher the impedance of the probe, the less effect it has on the circuit you're observing.

Figure 2-2. The oscilloscope probe becomes part of the circuit you're measuring, so it is important that it is a quality part

Oscilloscope inputs have capacitance associated with them as well. This may vary from scope to scope, but it is generally in the range of about 10 to 30 picofarads in X1 mode and one-tenth that in X10 mode (another X10 advantage). When measuring low level signals, it is best to use X1 mode if possible; otherwise, you're putting a X10 attenuator in front of the scope designer's quiet little input amplifier, and you may need the added resolution the X1 mode provides.

Last but not least, remember to *compensate* the probe in X10 mode to ensure that the impedance between the probe and scope input are matched; otherwise, you'll be observing a distorted waveform. Scopes generally have a square wave generator with a connection point on the face that provides a convenient compensation waveform. Just connect the probe (set to X10 mode) and adjust the variable capacitor on the probe until the wave looks square and not differentiated (front edge of square wave higher than normal) or integrated (front edge of square wave lower than normal). Use the plastic adjustment

tool that comes with the probe, as a metal one could affect the adjustment.

If you get a scope capable of at least 5mV per division on the low end, at least 5V per division on the high end, and X10 probe capability, you'll be in great shape. Higher priced scopes will have more sensitive input circuits and may permit higher voltage measurements as well.

What Else Should I Look for in a Scope?

When looking at an analog scope — either new or used — check for these additional features:

- Beam Intensity and Focus controls
- Vertical Magnification Factor button on Input Channels
- Sweep Magnification Factor button in Time/Div control area
- Position controls for Input Channels and Sweep Start
- Automatic Triggering and Trigger Level, Mode, and Source selection
- Variable adjustment controls for Input Level and Sweep Time
- Ability to invert one or both input signals
- Ability to do X/Y waveform display

Buying a scope with some of these features missing is not the end of the world, but the more of them you get, the happier you'll be with it. I heartily recommend that you try out a used scope before buying it, although that may be impossible if you're at a yard sale or flea market. The older an analog scope is, the dimmer the display becomes, and the more out of calibration it may be. As you progress in electronics, try to find a way to get access to a better scope if the first one you buy lacks essential features.

The Digital Multimeter (DM)

I'm only going to go over digital multimeters because they have become so inexpensive and ubiquitous (Figure 2-3 shows a typical digital multimeter). If you already have an old analog multimeter with a precision movement that provides you with enough accuracy, stick with it. However if you're just getting started, go for the digital multimeter. You can get a decent starter unit for the price of lunch at a fast food restaurant.

As the number of display digits or resolution goes up, multimeters get more and more expen-

sive. Another factor in multimeter cost is whether the unit is battery powered (lower cost) or includes a power supply (higher cost), as many bench models do.

One thing to keep in mind when reading multimeter specs is that in addition to the accuracy percentage, you will see an additional +/- least significant digit(s) count. This just means that the reading may bobble around a little in the least significant (or two least significant) digits, so the lower this range, is the better.

Always use the most appropriate range (if the multimeter does not auto-range) for your meas-

What's Special About Digital Scopes?

I have both a 40 MHz analog scope and a 50 MHz digital scope on my bench, and I use them both all the time. The features that set digital scopes apart from their analog couterparts are legion. The scope's control panel has many of the same controls as the analog scope for convenience. If you had to enter a special settings mode and scroll through a software menu just to set a channel's volts per division or the scope's sweep speed you would get *real* tired of it *real* fast.

However there are a *lot* of functions that *are* hidden behind menus that you access via general purpose *assignable* buttons typically located on the front panel next to the main display. In menu mode, the scope will show you a menu of functions on the screen next to these buttons and you use them in conjunction with a digital pot or two to select and adjust the parameter of interest. Basically a digital scope is an analog scope with really fast A to D converters for the input channels and a *powerful computer* to process the data streaming from the A to D converters. I'm not going to go over *every* function a digital scope provides, but here are some of the cooler features:

- High-contrast multicolor LED or LCD display

- Cursor measurements of both time and amplitude for the displayed waveforms

- Auto-measurements of rise time, fall time, frequency, duty cycle, etc. of the displayed waveforms

- Math functions such as fast Fourier transform (FFT)

- Pass/fail testing of waveforms

- Waveform image acquisition

- Scope "set up" memory and storage

- USB output for storing scope data and waveform images on flash drive

- Auto-calibration

- Capture and hold transient signals for display and measurment

There are times when, even with all these cool features available on my digital scope, I still find my analog scope very helpful. When you start working with a digital scope you'll see what I mean. Sometimes the digital scope picks up *too much* data and you find yourself looking at a lot of noise and digital artifacts that a typical analog scope CRT smooths over for you. In their defense digital scopes provide a variety of filtering functions to deal with this but digging through menus to find these functions is not always conducive to productivity in my opinion.

The bottom line when buying a digital scope is to go for the largest display, highest sample rate (for example 1G samples/sec), largest memory (for example 1M points), and most bits of resolution (8 at least) you can afford. Take care of your scope and it will serve you for years and years.

Figure 2-3. *A typical digital multimeter*

urement to ensure you're getting the most accurate reading. Setting your meter to the 2V range and trying to read 20mV is asking for inaccuracy. A more appropriate range would be the 200mV range. Set the meter so that the expected measurement is near the top of or at least well within the selected range. When measuring an unknown voltage set the meter to the correct mode (AC or DC) and highest voltage setting before adjusting the range down as appropriate.

What should the multimeter you buy be able to do? Here is what I consider to be the basic functionality and accuracy required for getting started in analog electronics work:

- Minimum three and one-half digits
- Minimum 2000 count precision
- 200mV or lower minimum DC voltage scale
- High-contrast readable display

- Measures voltage (AC and DC), current, and resistance
- DC voltage measurement accuracy should be at least +/−0.5% of full scale reading
- AC voltage, DC *current*, and resistance measurements should be accurate to at least +/− 1.0% of full scale reading

The more precise and accurate a multimeter is, the more expensive it's going to be. Low noise parts must be used, temperature compensation must be designed in, and the circuitry has to be reliable for years. The rather logical law of test equipment prices is this: *premium parts equal premium price*.

Multimeter Advanced Features

If you can afford to step up to a meter with these additional features, you'll be glad you did:

- Higher accuracy than listed above (.05%, .03%, etc.)
- Auto-ranging
- AC current measuring function
- Frequency counting function
- Capacitance measuring function
- Diode and transistor test capability
- Data hold function
- Delta measurement mode
- USB, GPIB, and/or RS232 computer interface

Logic Probe

A logic probe is a very simple piece of test equipment used to verify the state of a logic gate's output (Figure 2-4). Essentially, it is a pen-like device with two (red and green) or three (red, green, and yellow) LEDs that glow individually, depending on what logic level the probe's tip is contacting.

Figure 2-4. *A typical logic probe*

ways convenient to go tear down a patch you may have in progress, disconnect your synth-cabinet, bring it into your shop, and run jumper cables from its power supply to test something you're developing or troubleshooting. After you go through that rigmarole a few times, I'll bet buying a bench power supply moves higher up on your priority list. What should you look for in a bench supply?

Figure 2-5. *A typical bench power supply*

The logic probe must be powered from the circuit under test, and so it has two leads (typically red and black) that come out of one end and connect to the positive and ground of your circuit. When you probe the output of a logic gate, the logic probe typically lights the red LED for a high logic level, the green LED for a low logic level, and the yellow LED for an indeterminate logic level.

Another feature some logic probes have is pulse detection, which stretches out narrow logic pulses (both high going and low going), giving you a visual indication that a logic pulse was detected at the tip, even if it would normally be too narrow to detect visually. You can build a simple logic probe yourself or buy one with more features to save the trouble. Logic probes come in very handy for troubleshooting complementary metal-oxide semiconductor (CMOS) logic circuits in which levels do not change very rapidly. For observing rapidly changing logic signals, an oscilloscope is required.

The Bench Power Supply

When you are just getting started in electronics, you can work entirely with batteries. Eventually, however, you may want to expand your projects until batteries are no longer practical and a line-powered supply becomes necessary (Figure 2-5). When you're working on the bench, it's not al-

Most synth-DIY projects use a bipolar (or dual) DC power supply. Op amps, for instance, one of the most commonly used synth-DIY components, require a positive voltage, a negative voltage, and a ground. In synthesizers, the dual or bipolar power supply voltage is usually between 9V and 15V. Thus, the synthesizer's power supply might provide +9V, –9V, and ground; or +12V, –12V, and ground; or +15V, –15V, and ground, depending on the designer's choice. Occasionally you'll find a synth that has a bipolar supply as well as an additional +5V supply that is used to power logic chips or microprocessors.

What Is a Dual Power Supply?

It may not be obvious to people just getting started in synth-DIY, but most analog synth circuits require positive voltage, negative voltage, and a ground. The op amps in the synth's modules require this type of power to work properly. You can power op amps with a single voltage supply if you create a "virtual ground." To do this, use a resistor divider to create a voltage level that is halfway between the negative and positive poles of the single voltage supply. A capacitor is often added between the "virtual ground" potential and the supply's negative pole for stabilization (as seen in **Figure 2-6***). This "virtual ground" potential is then used as the circuit's signal ground. It should be noted, however, that a virtual ground is nowhere near as solid or low impedance as a "real" ground provided by a line-powered dual power supply.*

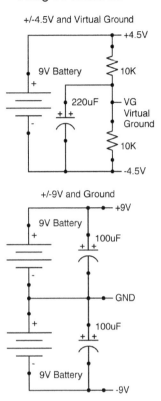

Bipolar Power Supplies
Using 9V Batteries

Figure 2-6. *A dual power supply can also be made from a single 9V battery by creating a virtual ground, or by using two 9V batteries wired as shown here.*

The upshot of all this is that you'll need at least a dual output power supply. A dual output power supply is a box with two independent power supplies whose output voltages can be adjusted individually. Often they include built-in voltage and current meters to display both the voltage setting and the current being drawn from the supply. More expensive units include a current limit setting that protects the circuit being powered from being destroyed by excessive current.

If you're just getting started, I recommend using batteries to power your electronic experiments. It's better to make your mistakes using batteries because they are less likely to damage a miswired experiment than a high current output bench supply. Figure 2-6 shows two ways to make a bipolar supply using 9V batteries and a few components.

About ATX Supplies as Bench Power Supplies

Be careful if you decide to harvest a power supply out of an old computer. Not only do they supply enough current to weld with (don't short the output without goggles on) but often the power is noisy, since they are **switching** *power supplies. A full-wave rectified dual-analog supply delivering a couple of amps of clean power is a better choice.*

The plus and minus connection points of both supplies are normally isolated electrically from the unit's case, which is connected to earth ground. Don't connect the positive or negative connection points to the unit's earth ground terminals when using the supply to provide plus and minus voltage to a circuit under test. First set the voltages on both supplies to the desired voltage level. For our explanation, we'll work with +12V. Connect the negative connection of one supply to the positive connection of the other supply. This will become the neutral (or signal ground) connection output point of the dual DC supply. The remaining unconnected positive terminal becomes the positive output of the dual DC supply (+12V), and the remaining unconnected negative terminal of the other supply becomes the negative output of the dual DC supply (–12V).

If you place the black lead of your DM on the junction of the connected positive and negative terminals and the red lead on the remaining positive supply output, you will measure +12V. Leave the DM's black lead where it is and place the red lead on the remaining negative supply output, and you will measure –12V. The dual bench supply you buy should have, at least, these basic features:

- Two independent power supplies, both adjustable between 0V and 30V
- Both outputs minimum 2A capacity
- Voltage and current meters
- Low output ripple voltage

If you can afford to step up to these additional features, you'll be glad you did:

- Current limit setting on both supplies
- Tracking capability (one supply's controls affects both supplies)
- Third output (5V at 2 or 3 amps)

Look around the major Internet shopping sites for *dual power supply*, and buy one with the features and price point you want. Use your favorite search engine to look for *electronic instrument*

suppliers to find even more distributors to check out.

Some Nice-to-Haves

Here are some instruments that, as this section's name implies, are *nice to have*. However, before you go out and buy them, look for a multimeter with more accuracy and functionality. I say this because you can save money by getting a multimeter that also measures capacitance and frequency. A multimeter of this kind will give you the ability to match capacitors or read frequencies when you need to, and you don't have to crowd up your bench with a bunch of instruments you only use once in a while. Often you'll want (or need) to measure more than one thing at a time, which is why I list a second multimeter in the *nice-to-haves*.

Second Multimeter

Making more than one measurement at a time comes up so often in electronics work that it just makes sense to splurge for another multimeter as soon as you can.

Capacitance Meter

A dedicated capacitance meter is nice to have when you need to match capacitors or check the value of a cap removed from equipment whose value may be marred or gone (Figure 2-7). They are generally auto-ranging, can measure a wide range of capacitance (example .1pF to 199.99 mF), have data hold, and include multiple measurement modes. As with multimeters, look for a high "count" unit, since the higher the count, the more accurate the readings will be.

Figure 2-7. *A typical capacitance meter*

However, unless you *really* need extreme measurement range and accuracy, go for a more functional multimeter with capacitance measuring capability.

Frequency Counter

Before I even get started, I repeat: *buy a multimeter with frequency-counting functionality for your first instrument.* If you know you're going to be making simultaneous voltage and frequency measurements, then buy a second multimeter with frequency measuring capability. Frequency counters are cool, but many of them take more time to "count" the frequency than you may think, which can be irritating when you need to make quick measurements in real time.

Many frequency counters have precise settable gate times, during which the number of transitions that take place on the input are counted (Figure 2-8). So if the meter reports that 10 transitions took place during a 1-second gate time, we know that the frequency is about 10 Hz. If the meter reports that 10 transitions took place during a 10-millisecond gate time, then we know the frequency is about 1 kHz. So if you want to look at a frequency that is below 1 Hz, get ready to wait while the counter counts the transitions

during the successive *10-second* gate times. The prize for your patience is 100mS accuracy (if you can call that accurate). Most dedicated frequency counters are mainly for counting high frequencies in the MHz and GHz region and give short shrift to the audio and lower-frequency range.

Figure 2-8. *A typical frequency counter*

The dedicated frequency counter certainly has its place, but a much faster algorithm for determining the frequency of a regular non-modulated signal is to measure its period accurately and then display the period's reciprocal value (1/(cycle period) = frequency in cycles per second). The more accurately you measure the period, the more precise the frequency measurement. Many multimeters do this, and in my lab it's the only way I measure frequency anymore. My dedicated frequency counter sits and collects dust. As Marley said to Scrooge: "Mark me… buy a multimeter with frequency counting functionality"—or something along those lines.

Function Generator

A function generator comes in handy all the time when working on analog synthesizers (Figure 2-9). I still list it as a nice-to-have because a hobbyist can make a simple but useful function generator for bench use. Since the waveform, amplitude, DC offset, and frequency of the output of the function generator can be adjusted, you can create whatever type of signal you need to exercise a module or circuit you're working on. Many function generators have CMOS and TTL (transistor–transistor logic) level signal outputs in addition to the normal waveform outputs, which also come in handy on occasion.

Figure 2-9. *A typical function generator*

Function generator output waveforms typically include sine, square, and triangle waves. Make sure the function generator you buy has adjustable waveform skew (or duty cycle), which gives it the ability to produce pulse, ramp, and sawtooth waves as well. Another cool feature is *frequency sweep* capability. This is the ability to repeatedly sweep the function generator's frequency using either a log or linear curve from a low frequency to a high frequency, with both sweep width and sweep rate being adjustable. This function is often useful when looking at the response of a filter or an amplifier to determine its bandwidth. As you feed the sweeping frequency into the circuit under test input, you observe the amplitude of the output of the circuit under test using an oscilloscope. The frequency at which the observed output signal's amplitude is 3dB lower (approximately 71% of the value being applied to the circuit under test input) is considered the top end of the circuit under test's *bandwidth*. Better (more expensive) function generators go higher in frequency, have more waveforms, and allow more precise sweep settings. If you can manage to add one to your instrument ensemble, then by all means do so, as you will find it useful.

Tips for Reliable Soldering

People often write to me regarding problems they're experiencing while getting a synth module to work properly. I usually encourage them

to find a friend or relative with electronics experience but also suggest they try this or that, depending on the situation. Nine times out of ten, the problem was a *bad solder joint*.

Reliable solder joints are the cornerstone of a reliable circuit. Improper solder joints can lead to all kinds of problems: intermittent behavior, noise, crackling, and even circuit failure. A host of factors affect the quality of solder joints. Here is a list of some of the important factors affecting solder joint quality:

- The temperature of the soldering iron
- The condition of the solder iron's tip
- The type of solder and flux used
- The cleanliness of the surfaces to be soldered
- The skill of the person soldering

Temperature-Controlled Solder Station

I must have gone through 12 dozen cheap pencil soldering irons before I finally broke down and bought a temperature-controlled soldering station (Figure 2-10). While the cheap pencil solder irons are an inexpensive way to get started, I heartily recommend that you save up and get an adjustable temperature-controlled solder station ASAP. I have seen units on Amazon selling for between $20 and $100, so it's not a major investment. I have a couple of soldering stations: one by a "name brand" company, and one by a "never heard of them" company. I use both of them all of the time and they both do the job just fine. So what exactly should you look for in a temperature-controlled solder station?

- Adjustable temperature range from 250 degrees C (500 degrees F) to 450 degrees C (850 degrees F)
- Working temperature attained within a minute
- Solder pencil holder included

- Tip cleaner sponge or brass coils included
- Soldering pencil that's comfortable to hold

Figure 2-10. *Typical temperature-controlled soldering station*

Alright, now that we have a decent solder station, how do we put it to the best use?

Soldering Tips

I recommend 1/32-inch thick, rosin-cored, 60/40 (tin-lead) solder. When wiring panel components, you might want to use a thicker solder (1/16 inch) of the same type. Buying solder in bulk is less expensive than buying small rolls or tubes. The one-pound spool of solder, while a bit of an investment, is actually the most cost-effective way to go and will last a long time.

Set the temperature of your solder station to between 320 degrees C (608 degrees F) and 380 degrees C (716 degrees F). The 60/40 (tin-lead) solder melts at a lower temperature than that (215 degrees C or 419 degrees F), but when you apply the soldering iron's tip to whatever you're soldering, it rapidly loses heat. This necessitates the higher temperature setting to ensure that the solder becomes molten quickly and good eutectic bonding takes place.

The rosin flux in the solder has several important benefits. It prevents surface oxidation during heating, cleans the surfaces to be soldered, and lowers the surface tension of the molten solder, helping it to flow readily and form a reliable eutectic bond between itself and the items being soldered. Always remove rosin flux residue after soldering with an appropriate solvent for the rosin type. Choose an environmentally friendly type of flux remover (no chlorinated fluorocarbons) to keep the ozone layer and your karma in good shape. Don't apply power to the PC board until you have removed all of the cleaning solvents and the board is dry.

Get yourself a chemical pump bottle and keep it full of a mixture of 75% isopropyl alcohol and 25% acetone, and you'll have a great solvent for cleaning rosin flux. If you can get the 99.9% pure isopropyl alcohol (drugstore variety is 91% pure), the mixture will contain less water, which can be absorbed by PC boards and components, but don't obsess, since the 91% variety seems to work just fine. Use solvent-soaked plastic or horse hair bristle brushes to scrub areas in need of flux removal. I like to cut the bristles short on a few brushes for when more scrubbing power is needed. Cotton swabs can be used for cleaning but can leave fibers behind. Canned air is good for blowing the excess solvent and suspended flux off of the board as well as drying it after cleaning. When the solder joints are shiny and the PC board material looks clean, not dull and streaky, you're good to go. Figure 2-11 shows the basic tools necessary for reliable soldering and desoldering.

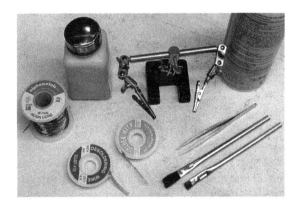

Figure 2-11. *Tools you'll need for soldering*

Setting the tip temperature too high (not far above 380 degrees C) is just bad all over. You will evaporate the solder's rosin core, preventing it from doing its important work. You will melt plastic elements of components that adjoin terminals. You will delaminate pads from your PC board. You may cause value changes or worse, destroy temperature-sensitive components. Stay within the safe temperature range (320 to 380 degrees C), and don't linger on a soldered joint once the solder flows.

Be careful when soldering switches, jacks, or any components that have plastic parts. Applying the solder pencil's tip to a component's terminals for too long or with too much pressure or heat can transfer heat to any surrounding plastic, melting it and damaging or destroying the component. Turn the heat down a bit when you're about to solder components with plastic parts. Experiment on a spare part so you know how it will react when exposed to soldering temperatures.

Rosin Paste Flux

I've mentioned that excessive heat can damage plastic parts of panel components. To minimize the solder iron application time, I apply a **tiny** *amount of Qualitek 50-4002B PF400 Rosin Paste Flux with a toothpick to terminals prior to soldering. The additional rosin flux makes the solder wet to the panel component terminals more quickly, reducing their exposure to high temperature.*

Let the soldering iron heat up before beginning. Dampen the solder station's cleaning sponge by first wetting it and then squeezing the excess water (and any solder balls) out of it over the trash can. Damp is good but soaking wet is…not so good. You want the sponge to be able to clean the tip, but you don't want to cause the tip's temperature to fall excessively by dipping it into a

soaked sponge. After a few wipes on the damp sponge to remove any lingering old solder or oxidation, apply some fresh rosin core solder to the tip and wipe it on the sponge again. Keeping the tip coated with fresh solder will help it to transfer heat quickly to the items being soldered.

You don't have to clean every pad/lead combination with solvent prior to soldering because the rosin flux core in the solder does an admirable job of cleaning the surfaces. I can't remember the last time I was plagued with a cold solder joint, so history bears me out on this. Of course, if there is obvious oxidation on a component lead or pad, you should take the time to clean it before soldering.

When soldering a joint, you want to contact all of the surfaces to be soldered with the tip in order for heat to transfer to them evenly. Figure 2-12 illustrates the recommended application of the soldering iron tip to all surfaces for best heat transfer. After you apply the solder iron, immediately apply the solder so that it contacts both the tip and the heated surfaces. Once the solder flows, remove the tip and let the joint cool undisturbed. Don't linger after the solder has flowed, or the bad things discussed above will happen, possibly ruining a perfectly good day. Don't apply too much solder. The soldered joints should not look like bumps but should have concave fillets. Practice on a scrap PC board until you feel comfortable and your solder joints look shiny and have nice fillets.

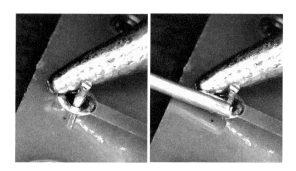

Figure 2-12. *Proper solder tip and solder application*

I generally solder about 10 to 16 component leads and then repeat the tip cleaning process. During soldering, if you apply the solder so that it contacts the tip and the materials being joined, you will ensure the tip has fresh solder and rosin on it, which will keep it from oxidizing. However, after you solder a number of joints, the tip will start to accumulate crud, and you'll need to clean it. With experience, you'll find the optimum number of solder joints you can make between tip cleanings.

Double-sided PC boards have what are known as *plated through-holes*. During the manufacturing of PC boards, the holes for mounting components are drilled through the copper-clad material first, and then those holes are plated through with copper (Figure 2-13 shows a cross-sectional view of a plated through-hole). This is a process that causes copper to be plated onto the walls of the component mounting holes drilled through the PC board material. These plated through connections give PCB pads extra strength and resistance to delamination and are used to route electrical signals between the two sides of the PC board. AAfter plating the insides of the through-holes with copper, the etching, solder coating, and solder masking processes are completed.

Figure 2-13. *Plated through-hole*

When soldering a double-sided PC board, the solder flows onto the donut pad surface and component lead, through the cavity between the component lead and the sides of the plated through-hole, and all the way up onto the top side's anular ring (Figure 2-14). This is exactly what you want, and you'll see solder fillets (metallic concave surface surrounding the component lead) formed on both the bottom and the top of the PC board.

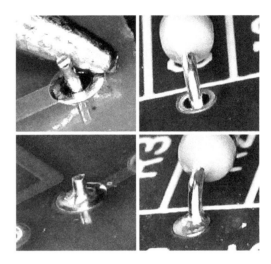

Figure 2-14. *Solder flow through plated through-hole*

If too much heat is applied by the tip and the rosin boils away, you may get poor wetting, the solder may appear bumpy and gray, and no top fillet will be seen. STOP! Turn the heat down a bit and apply some fresh solder (with rosin in it) to *reflow* the joint. A good solder joint should appear bright and shiny, especially after cleaning (Figure 2-15). Too much solder applied to a joint or that accidentally bridged two pads may need to be removed using desoldering braid (discussed below).

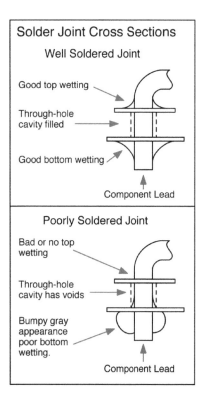

Figure 2-15. *A well-soldered joint versus a poorly soldered joint*

Desoldering Is Important, Too

You've been troubleshooting a circuit, and lo and behold, you find a component you need to remove. Here is what to do. On the top side of the PC board, cut the component's leads and discard it. Turn the board over. Using aluminum or stainless steel tweezers with a fine point, grasp the remaining lead with one hand and apply the soldering iron tip so that it contacts the PCB pad and the lead simultaneously. When the solder becomes molten, *gently* pull the lead with the tweezers and remove it. If the solder solidifies, apply a bit more solder (we want its rosin flux) and try again. The lead should come out with little effort. Once you've gotten the leads out, it's time to clear the holes of solder so you can install the replacement component.

I find solder braid extremely useful, so that's what I'm going to discuss. I think spring-loaded vacuum solder suckers have too much capacity to do harm to circuit board pads, so I very rarely use them. I will occasionally use a solder sucker for desoldering jack or pot terminals (things I can't suck pads off of), but thicker braid works fine for that, too.

Desoldering braid is made of very light gauge copper wires that are braided and impregnated with rosin flux. It comes in a variety of widths, and the wider it is, the more capacity to *absorb* or *wick* solder it has. Desoldering braid is not very expensive, so buy a few rolls of varying widths to have on hand. Figure 2-16 shows some typical solder wick dispensers.

Companies make special vises that can hold a PC board during soldering. They are very convenient to use but in no way required. When populating a PC board for soldering, your best bet is to install a few components at a time onto the board. Bend the leads protruding through the bottom of the PC board so the components don't fall out of the board when you invert it to trim them. In higher density areas, bend the leads away from any nearby pads to avoid accidental solder bridges from forming.

The reason I suggest trimming component leads prior to soldering is that the physics of trimming the leads post-soldering can apply a significant shock to the solder joint that can break the joint's eutectic bond. Something like that has the potential to haunt your circuit's future reliability, so before soldering, always trim the leads using a lead trimmer designed to reduce mechanical shock to the component.

Figure 2-16. *Two widths of desoldering wick*

To use desoldering braid, expose a couple of inches of new braid and then use the plastic case as a handle. Don't hold the copper braid itself, as it can get quite hot. We're going to apply the braid to the pad whose hole we want to clear and then apply the soldering iron's tip to the braid. If possible, apply the braid so that an eighth of an inch or so extends beyond the pad. Always clean and re-tin your soldering iron's tip immediately prior to applying it to the braid. Apply the cleaned, newly tinned tip to the braid right over the pad to desolder with light force so that the heat from the tip transfers to the braid, which in turn transfers to the solder on the pad and melts it. You should see the copper-colored braid turn silver as the solder flows into it. In Figure 2-17, I have removed the soldering iron but left the braid for you to see the silver color that forms as the solder flows into the braid. In practice, as soon as you see the braid become silver, remove it and the soldering iron tip simultaneously. Nine times out of ten, the hole will be clear but if it's not, don't panic; here's what to do.

Figure 2-17. *Clearing a pad's hole with desoldering braid*

Apply a bit of fresh rosin core solder to the pad's hole and try again. Sometimes you need to *prime* the braid with a tiny bit of solder to get the heat to transfer properly. Don't use too much, or the braid will become saturated, rendering it incapable of wicking additional solder. With practice, you'll hit the nine-times-out-of-ten mark and only have to fuss with a pad once in a while.

Obtaining Electronic Components

I've been buying electronic components for decades, and one thing I've seen over the years is that prices are all over the map. Always shop around when you're preparing to gather parts for a project. Potentiometers, switches, resistors, capacitors—any electronic component can be purchased online from hundreds if not thousands of vendors. I'm not saying that you should abandon your local electronics retail store, especially if it provides a wide selection of parts at competitive prices. However, if that has been the *only* place you've been buying components, it's time to find out what online electronics stores have to offer.

Surplus Parts

You'll want to stock your *lab*, be that the desk in your bedroom or a space set aside in your attic, basement, or garage. When it comes to things like resistors and capacitors, you can often find great deals by scouring the surplus sites. Use your favorite search engine to look for "surplus electronics," and you'll be visiting websites for as long as you've got time.

Capacitors

Often, surplus electronics sites will sell capacitors that they've bought from companies that have overstock or are going out of business. If you can't find the value you need there, remember that you can create capacitance values you don't have in your stock by putting caps in series or parallel with one another (see Figure 2-18).

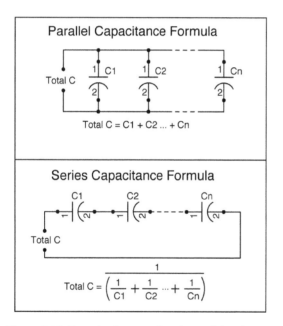

Figure 2-18. *Formulae for capacitors in parallel and series*

There are several types of capacitors commonly used in analog synths: nonpolarized ceramic and film caps, nonpolarized aluminum caps, polarized aluminum electrolytic caps, and polarized tantalum electrolytic caps. The polarized types must be connected so that the polarity of the applied voltage matches the polarity markings on the body of the cap. Applying reverse voltage to a polarized cap will damage or destroy the capacitor, rendering it leaky or otherwise physically compromised.

Ceramic caps are great for coupling AC signals within a circuit where DC blocking is important. Film capacitors have characteristics that come in handy in certain applications. For example, polycarbonate and polystyrene film capacitors have great temperature tolerance and very low leakage. They are often used in sample-and-hold circuits where the capacitor has to hold a voltage for a long time without drooping.

Tantalum electrolytic capacitors have very low leakage and are great in instances where you're putting a large resistance in front of the cap and charging it. In this circumstance, if the cap exhibited excessive leakage, it might, for example, never charge to a necessary threshold voltage. To appreciate what capacitor leakage is like, just imagine putting a 10M or lower resistor in parallel with the cap you're trying to charge. The leakage is always trying to discharge the cap whether you want it to or not.

Aluminum electrolytic capacitors come in values from submicrofarad to several farads (yep, I said farads). They are often used for localized stabilization of voltage levels on PC boards in the 10µF to 100µF range and come in values large enough (tens of thousands of µF) to be used as the main filtering caps in power supplies. They are useful in timing circuits, filters, AC coupling, power supplies, and many, many other uses. Aluminum electrolytic capacitors have higher leakage than tantalum types.

Bipolar electrolytic caps provide a high value of capacitance, without the polarization normally associated with electrolytic caps. They use a trick you can use when you need a large amount of capacitance but don't want polarization issues. If you connect the negative poles of two polarized caps, you can use the remaining two positive poles as the leads of a bipolar (or nonpolarized) cap. The value of the two caps in series can be determined from the formula shown in Figure 2-18. If the caps have the same value, 10µF for example, the value of the newly formed bi-polar cap will always be one-half of the original values, in this case 5µF.

The *working voltage* of a capacitor is the voltage that can safely be applied to its plates before the dielectric material between them breaks down, resulting in destructive arcing. To avoid this, always buy capacitors for your project whose working voltage is well above the highest voltage you plan to apply to them in your application. Normally this is the power supply voltage, but a safety factor of 1.5 to 2 times the power supply voltage will keep your project's reliability up. Thus, if you are working on a project in which your caps might be exposed to a potential of 20V, then buy caps with a working voltage of 35V or more. For a 9V battery-powered project, 16V or 25V caps would fill the bill.

The physical size and lead spacing of capacitors can be important. Film and ceramic capacitors with lead spacing of 5mm (.2") are very convenient for breadboard work. Getting caps with too high a working voltage can result in caps that take up more physical space than necessary for a lower voltage project, so keep package size in mind when buying caps. Capacitor data sheets contain detailed size specifications, and you should become familiar with them.

For people who are just getting started, I suggest building projects that come as kits to ensure that you have all of the correct parts. This not only provides a more satisfying project experience, but it also teaches you what the components look like and what size and type to buy next time. Sourcing the parts yourself for projects is definitely the most cost-effective way to go. Request catalogs from the larger online electronics part distributors and look through them to familiarize yourself with what parts are out there, what sizes they come in, and what they cost.

Any time I've been in an artist's studio, I've noticed that there are normally several tubes of the colors used, as well as an easel, a palette, brushes, and canvases ready to go for when inspiration strikes. Having to stop and round everything up can be a real creativity killer. As you progress, you'll find that having a stock of commonly used electronic components on hand is practically a necessity. That way, when a circuit idea pops into your head, you'll be able to plunk it onto a solderless breadboard and experiment with it, which is one of the best ways to learn. What part values should you have on hand?

Capacitors to Keep On Hand

I suggest that you stock your lab with these values of capacitance for your experiments. Buy in bulk to save money. You'll notice that I've gone with an approximate 1, 5, 10, 50... series of values. Again, bear in mind that you can put these values together in series and parallel for additional values. When opportunity knocks, I suggest you expand your parts cabinet to contain a 1, 2, 5, 10... series of values. A series like that would go 10pF, 22pF, 47pF, 100pF, 220pF, 470pF, etc. You can generally find surplus ceramic or film caps that have these values, and for experiments, either type will work fine. Capacitor assortments are available that contain a nice range of values for breadboard experiments. You can buy tantalum or aluminum electrolytics for values at and above 1µF.

Value	Max Voltage Rating	Type
10pF	50V	Ceramic Capacitor
47pF	50V	Ceramic Capacitor
100pF	50V	Ceramic Capacitor
470pF	50V	Ceramic Capacitor
.001µF	50V	Ceramic Capacitor
.0047µF	50V	Ceramic Capacitor
.005µF	50V	Polystyrene Capacitor
.01µF	50V	Ceramic Capacitor
.047µF	50V	Ceramic Capacitor
.1µF	50V	Ceramic Capacitor
1µF	35V	Aluminum Electrolytic Capacitor

4.7µF	35V		Aluminum Electrolytic Capacitor
10µF	35V		Aluminum Electrolytic Capacitor
47µF	35V		Aluminum Electrolytic Capacitor

The .005µF polystyrene capacitor is handy for VCO and sample and hold experiments.

Resistors to Keep On Hand

Surplus houses will often offer new resistors sold in either tape and reel format or bulk. It's way cheaper to buy resistors by the hundreds (or more if you use enough of them) than to buy them one at a time, so take advantage of bulk pricing and stock your lab with the values I suggest below. As in the case of capacitors, resistor experimenter sets with 5 or 10 pieces of a wide range of common resistance values are also available. It's good to have one of those on hand as well as a good number (20 to 100 each) of each of the following values for your breadboard experimentation work.

Value	Max Power Rating	Accuracy	Type
100 ohm	1/4W	5%	Carbon Film or Composition
1K	1/4W	5%	Carbon Film or Composition
2K	1/4W	5%	Carbon Film or Composition
4.7K	1/4W	5%	Carbon Film or Composition
7.5K	1/4W	5%	Carbon Film or Composition
10K	1/4W	5%	Carbon Film or Composition
20K	1/4W	5%	Carbon Film or Composition
47K	1/4W	5%	Carbon Film or Composition
75K	1/4W	5%	Carbon Film or Composition
100K	1/4W	5%	Carbon Film or Composition
200K	1/4W	5%	Carbon Film or Composition
470K	1/4W	5%	Carbon Film or Composition
1M	1/4W	5%	Carbon Film or Composition
2M	1/4W	5%	Carbon Film or Composition
4.7M	1/4W	5%	Carbon Film or Composition
10M	1/4W	5%	Carbon Film or Composition
100K	1/4W	1%	Metal Film
200K	1/4W	1%	Metal Film

Resistors can also be arranged in series and parallel to obtain values you don't have in your parts cabinet. The formulae for determining the resistance of resistors in series and parallel are shown in Figure 2-19. The 1% resistor value parts drawer suggestions are for times when you want more accuracy, say for a D to A converter or a precision gain circuit.

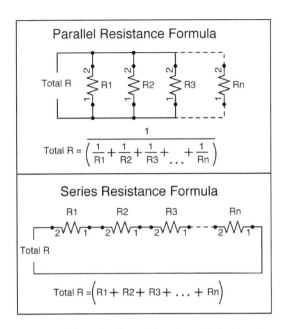

Figure 2-19. *Formulae for resistors in parallel and series*

Mechanical Components

Mechanical parts like potentiometers, battery snaps, switches, transformers, wire, etc., are often available from surplus houses, and buying them there can save you a lot of money. Revisit the same places you found when doing your Internet searches for "electronic surplus," and look for pots, switches, and transformers specifically to see what they've got. When all else fails and you can't find what you're looking for on surplus sites, you'll have to bite the bullet and go to a bona fide electronics distributor like *Mouser*, *Digi-Key*, *Jameco*, *Avnet*, etc. Compare who has the best price and buy it there. Do an Internet search for any of these distributor names to quickly find links to their websites.

Potentiometers are a common feature in most analog synths, and prices for them can vary a lot. A typical potentiometer can cost between $1 and $50, and if you don't know what you're looking for, you may spend way more than you need to. You'll typically want to buy carbon film or conductive plastic pots with a power rating between 300mW and 500mW, tolerance of 20%, panel-mounted configuration, in a 24mm or 16mm case size. Pots come with a variety of shaft lengths and types. There are knurled shafts, "D" shafts, and round shafts.

To avoid ending up with some goofy-looking, long-shafted pots, *always* check part numbers and specs before clicking OK. When in doubt, call the company you're buying from and ask for help selecting the correct pot/shaft/knob combination. You don't want to be up pot creek without a knob, to borrow a phrase. I buy a lot of the *Alpha Taiwan* brand potentiometers (which have a 1/4-inch-thick, round shaft and length of .335 inches) from a variety of part vendors; they work very well and are reasonably priced. Knobs for this type of pot are very common and come in a wide variety of shapes, sizes, and colors.

It's a good idea to keep some pots on hand for breadboard experimentation. I use solderless breadboards that work with 22 AWG solid wire when experimenting. For convenience, I've soldered six-inch lengths of 22 AWG solid wire to the terminals of several pots whose values I find useful and use them regularly during my experimentation. Keep five or so potentiometers of each of these values (10K Linear, 100K Linear, 1M Linear, 100K Audio, and 1M Audio) on hand for breadboarding. Other than that, I buy pots and knobs on a project-by-project basis, since it would be an expensive overkill to keep scores of pots on hand.

Switches are another part you need all the time and whose price can vary wildly. Take it from me that you can find decent-quality mini toggle switches for around a dollar each if you look hard enough. I've seen a mini toggle switch SPDT (single pole double throw) go for between $0.75 and $5.00 each. The generic mini toggle switches I buy that are in the $1.00 range are UL approved and have a current rating of 5A, which will handle any situation typically found in an analog synth. So look around, compare prices, and don't waste your money on name brands or excessively expensive components. Again, the surplus houses generally have all of the switch configurations you'll need (SPST, SPDT, DPDT, 3PDT, various push buttons, etc.). Do an Internet search for *mini toggle switch sales* or *push button sales* to find hundreds of distributors and switch styles.

For breadboarding, I've soldered six-inch lengths of solid 22 AWG wires to the terminals of a number of switches, just as I have for the potentiometers. It adds a lot of convenience to my experimentation not to have to stop and add wires to switches or pots while I'm in the heat of an experiment.

Transformers are something you run into when you start to build power supplies. You can buy power supplies prebuilt with multiple output voltages from several sources or make one yourself. If you go the *self* route, make sure you buy a transformer that has the correct primary AC voltage for your area of the world.

The most important thing after that is to get the right *secondary* AC voltage. When you rectify the output of a transformer, the rectified voltage is approximately the square root of 2 (1.414) times the VAC (Volts Alternating Current) value of the transformer's secondary (or secondary segment in the case of center-tapped transformers). Depending on the voltage regulator chip you use, you typically need to apply about three to five volts above the regulator's output voltage for proper operation of the voltage regulator. See the data sheet of the regulator you plan to use to determine what the secondary voltage of your transformer needs to be.

Center Tapped Transformer With Full Wave Bridge Rectifier

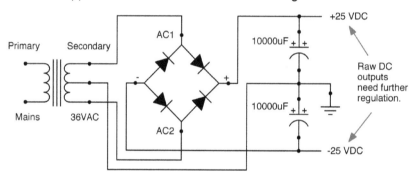

Figure 2-20. *Center-tapped transformer with full-wave bridge rectifier*

Most analog synths use a dual (also known as a bipolar) power supply, and they are most readily constructed using a *center-tapped secondary* transformer. For instance, a center-tapped 36VAC transformer has three wires on its secondary side. Two are attached to the two ends of the secondary coil, and the third is attached to the center of the secondary coil (the *center tap*). If you put the leads of your multimeter (set to measure AC voltage) across the *center tap* and either outer secondary wire, the meter will display 18VAC. If you put the multimeter's leads across the two outer secondary wires, you will read the full 36VAC. This arrangement is perfect for driving a

full-wave bridge rectifier to supply a beefy dual DC supply.

Figure 2-20 shows the skeleton for making a simple but effective dual DC power supply. All you need to do is add some voltage regulators to the raw DC outputs and voila—you have a regulated power supply with 1A or more of output capacity, depending on the regulators used. The LM78XX/LM79XX series or three terminal regulators are widely used and provide up to 1A or more of regulated power to your project. They come in a variety of fixed voltages (LM7805/7905 +/–5V, LM7809/7909 +/–9V, LM7812/7912 +/–12V, LM7815/7915 +/–15V, etc.). I highly suggest that you add a switch and a fuse to the primary side of any mains powered supply.

For adjustable supplies, you can use the LM317/ LM337 adjustable positive and negative voltage regulators. You could even make your own adjustable dual DC bench supply using these regulators. By reading the data sheets for the regulator chips (a skill you should definitely develop), you'll see exactly how easy they are to put into your project.

Active Components

Active components are things like integrated circuits, transistors, and diodes. There are hundreds of thousands (if not millions) of different active devices available for electronics. You will often need to buy specialized components for projects, and you can't stock everything, but

Table 2-1 lists some transistors and ICs to stock for your experimental work. Electronic parts come and go, so the specific integrated circuits I mention here may not be available in the future, which is why I will specify the function of each chip. Find the current version by searching online for the description rather than the specific part number, or try both. The *Music From Outer Space* website is a great resource for hard-to-find active and passive components.

This is my suggested list of active parts to stock in your parts cabinet. I have several of each of these devices on hand for experimentation. Look up the data sheets for the components you're not familiar with to discover the unique functions they provide. I'll cover specific uses for some of these devices in Chapters 4 and 5 as well as the appendixes.

Table 2-1. Active components parts drawers

Part Number	Function	Great For
2N3904	General Purpose NPN Transistor	Driving LEDs, discrete amplification, log converters, current sources, noise sources
2N3906	General Purpose PNP Transistor	Driving LEDs, discrete amplification, current sources
2N5457	N-Channel JFET Transistor	Source followers, sample and hold, integrator reset
PN4391	N-Channel JFET Transistor	Sample and hold, integrator reset
MPF102	N-Channel JFET Transistor	Sample and hold, integrator reset
LM13600/LM13700	Dual Transconductance Op Amp	VCF, VCA, VCO, and a million other circuits
LF444	Quad JFET Input Op Amp	Sample and hold circuits
TL074	Quad JFET Input Op Amp	Amplification, filters, comparators
TL072	Dual JFET Input Op Amp	Amplification, filters, comparators
TL071	JFET Input Op Amp	Amplification, filters, comparators
CD4066	CMOS Quad Analog Transmission Gate	Sample and hold, gating functions, effects
CD40193	CMOS 4-Bit Up/Down Counter	A to D, counting, timing and logic
CD40106	CMOS Hex Schmitt Trigger Inverter	Timing and logic
CD4094	CMOS 8-Stage Shift-and-Store Bus Register	Timing and logic
CD4013	CMOS Dual D Flip Flop	Frequency division, switch debouncing, data latching, timing and logic
CD4001	CMOS Quad NOR Gate	Logic and flip flops
CD4011	CMOS Quad NAND Gate	Logic and flip flops
CD4024	CMOS 7-Stage Ripple Carry Binary Counter	A to D, counting, timing and logic
LEDs	General Purpose LEDs (various colors)	Lighting up your project
1N914 or 1N4148	High-Speed Diodes	General rectification, timing and logic

Many online sources carry many or all of these devices. Some of the N-Channel JFET transistors are becoming harder to find but are still readily available at *Music From Outer Space* and other boutique synth-DIY sources. So what are we going to do with these components we've gathered? That's where the solderless breadboard comes in.

Solderless Breadboarding

Solderless breadboards (SBBs) are standard equipment for any aspiring synth-DIYer. They're often the place where new circuit ideas are born and honed. They're super handy for learning and/or experimenting. I can't think of an easier way to set up and tear down a circuit than using one of these guys.

Solderless breadboards are equipped with phosphor-bronze, nickel (or gold) plated wire-clips arranged on a one-tenth-inch grid. The one-tenth-inch spacing is the same as that of dual in-line package (DIP) integrated circuit leads. The clips are designed to accept component leads and solid hookup wire (22 AWG is a great size) and provide good electrical connections for low-frequency, low-current work.

Kits are available that contain a variety of wire lengths and colors with prestripped ends, prebent at 90 degrees for use with solderless breadboards. A less expensive alternative is to buy some spools of solid 22 AWG wire, cut it to length, and strip the ends as you work. Bending the ends at 90 degrees is not necessary, and you can just poke the stripped wire ends into the breadboard. Don't poke big fat leads into the breadboard or you may overbend the clips, reducing the reliability of the connection points. When I'm finished with a breadboard session, I gather up the stripped wires and keep them in a bin for next time. Figure 2-21 shows a typical solderless breadboard and two spools of 22 AWG solid hookup wire all ready to breadboard the next killer circuit.

Figure 2-21. *Solderless breadboard and solid 22 gauge hookup wire*

Figure 2-22 shows the typical connection scheme used on solderless breadboards. The red lines indicate which of the clips are connected together. The power buses on the sides of the SBB are connected along the length of the SBB. Beware that often the power bus is only connected halfway down the side of the SBB. You may need to add a jumper between the two segments of the power bus if you want power all the way down the strip. The connection points that go across are connected together but do not continue through the center of the board to the other side. They are separated specifically so you can plug a DIP integrated circuit with one-tenth-inch lead spacing into the SBB and then connect its leads via the remaining socket clips on either side of the chip to additional components mounted in other areas of the SBB (Figure 2-22).

Figure 2-22. *Solderless breadboard highlighting connected clips*

Hand Tools

In Figure 2-23 we see the typical set of handtools you'll need to get started. You'll want to buy a pair or two of fine-point tweezers for use in your soldering and desoldering work. I like the stainless steel types because they're strong and they resist corrosion. Aluminum tweezers are also good but less durable. An advantage of the aluminum tweezers is that they remove less heat during desoldering when the soldering iron tip comes into contact with them.

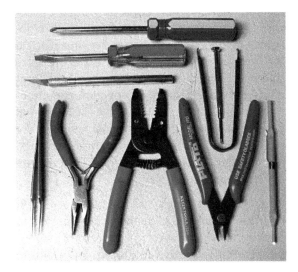

Figure 2-23. *Hand tools for synth-DIY*

You'll also need needle-nose pliers, a good pair of diagonal cutters, a wire stripper (with small AWG stripping capability), Phillips and flat bladed screwdrivers (including small jeweler's screwdrivers), an IC remover, hobby knife, and a trimpot adjustment tool. Also important to have on hand is an assortment of multicolored test leads with alligator clips or mini-clips on the ends. They're perfect for quickly connecting things on the work bench. You'll want to get yourself a static mat and wrist band with a connecting cord to aid in combating static damage to your active components. A nice magnifying glass is invaluable when trying to read the microwriting on small components or transistors. Lastly, a fluorescent lamp with a magnifier is invaluable for PC board inspection, soldering, desoldering, and troubleshooting.

Coming up with cases for your projects is part of the fun, and when you're just getting started, you can use available materials to save money. When I first got started, I used aluminum bake pans for making chassis and face plates. They're cheap, readily available, and can be drilled easily. When breadboarding and coming up with circuits, I'll often mount the panel controls on a piece of cardboard with HVAC aluminum tape on the back to provide a ground plane.

You can also buy premade aluminum and plastic cases from electronic suppliers to house your projects, but even then you'll need at least an electric drill to make the component mounting holes. If you plan to make your own cases for your projects, you'll need at the very least a power drill, a jigsaw, a vise, sandpaper, nails, and screws. Start small and obtain tools as you need them. You can find deals on used tools at garage sales and flea markets.

Well, you're all tooled up and you have the instruments you need to troubleshoot your projects. In the next section I'm going to cover some troubleshooting basics. Anyone building electronic equipment will tell you that troubleshooting your work is standard procedure, and the better you get at it, the faster you'll be enjoying the projects you make.

Troubleshooting Tips

You've populated your PC board with all of the components, soldered carefully, made a front panel, mounted the components on it, and wired it to your PC board. With high anticipation, you power it up, connect a cable from your project to your amp, and… nothing. To quote a famous guide often used by universal hitchhikers, "DON'T PANIC." It's not the end of the universe, you're not a failure, you're still every bit as special as your mom told you—you simply have a bug and need to do some troubleshooting.

You should put on a pair of goggles when powering up a project. There are components that will explode if they're not installed in the proper orientation. Tantalum caps, for instance, go off like firecrackers if installed backwards.

Everyone makes mistakes, especially when building an electronic project. I've been building electronic projects for decades, and I still goof from time to time. I forget a part, wire something to the wrong place, forget a wire entirely, or make some other simple mistake. The trick is knowing how to track down the problem and fix it, and that's what I'm going to try and prepare you to do.

How to Minimize Troubleshooting

You can cut way back on troubleshooting if you're ultra-careful and attentive to every detail during the construction of your project. Observe static precautions in your work area. Get yourself a static mat and a wrist band with a cord that connects to the mat. Wear it while working with active components (ICs and transistors) to ensure you don't statically damage your components while handling them. Use a magnifier light when inspecting your PC board's component values and solder joints.

I advise people to print out the PC diagram with the component legends or values on it so they can work from it. Then after the printed circuit board is populated, go back and check that you put the correct components in every location and that each of them is properly soldered. Inspect the resistor color bands carefully to be sure each is correct. For example, it's very easy to mistake a 200 ohm resistor (red/black/brown) with a 1K resistor (brown/black/red). Be certain to look carefully at the capacitor values to be sure they're all correct as well. Make sure all active components and IC sockets are oriented correctly. A transistor, diode, or IC that is oriented incorrectly will definitely cause havoc. Use IC sockets for all

ICs just in case you have to go looking for a bad one. It's way easier to desocket than to desolder.

Make sure there are no component legends without a component on the PC board. Look the board over very carefully until you are certain all components have correct values and are soldered correctly. If you find a component with an incorrect value, carefully desolder and replace it. If you see a suspect solder joint, reflow it (add a tad of fresh rosin core solder) or desolder it using desoldering braid and resolder it.

I also recommend that people print out the panel wiring details for a project and then use a highlighter to mark off each wire as they install it. Do the intra-panel wiring first. These are the wires that interconnect the components on the front panel. Here again, careful inspection and attention to detail will give you the best chance of success on power-up. Go over each wire in the wiring diagram, comparing it to the wiring you just installed. Install any missing wires or reroute any that got misrouted.

When you're ready to wire the panel to the PC board, look over the schematic for the project to determine which wires will be carrying which signals. This is another reason it's important to learn to read schematics well. Keep inputs and outputs away from one another. Keep high-level signals away from low-level signals or inputs. When you put two wires next to each other, parasitic capacitance and inductance exists between them, which can cause a high-level signal in one wire to be coupled into a separate but adjacent wire.

Examples of high-level signals include VCO outputs, LFO outputs, gate signals, trigger signals, and comparator outputs. Examples of sensitive signals susceptible to reflection from high-level signal-carrying wires include mixer inputs, filter inputs, and VCA inputs. Use coax cable for sensitive inputs and only ground one end of the cable to avoid ground loops. By routing your wires properly, you'll avoid issues related to signal reflection.

What to Look For

Here are a number of things to consider when trying to find an issue with a circuit. This is where your oscilloscope, multimeter, and Sherlockian sleuthing skills (Figure 2-24) are going to get a good workout.

Figure 2-24. *A loupe is handy for close inspection of components and solder joints*

- Read over the circuit's description so that you are thoroughly familiar with what it's *supposed* to be doing.

- Make a copy of the schematic and use it to record notes and measurements made during troubleshooting.

- If you are lucky enough to have another working version of the circuit, use it to compare what you are seeing in the defective one. This is one of the easiest ways to find a problem within a circuit.

- Look very closely at changes or modifications you may have made to the circuit, even if it worked fine for a while after the change was made. Consider any degradation in circuit operation observed before the circuit stopped working. Did you connect the input or output to something you hadn't connected it to before? Considering these things can give you a leg up on where to start looking for the problem.

- Go over the component values on the PC board again to make sure every one of them is properly oriented and of the correct value.

- Go over the wiring again, as this is the most common cause of errors. Forgetting any wire is problematic, but when it's a ground or power wire that subsequently connects to other panel components, you'll see erratic behavior.

- If an IC is hot, it may not be oriented correctly. Powering an IC with reverse voltage will absolutely burn it out and destroy it. You might as well replace any chips you installed backwards because *they're dead, Jim*.

- If you etched the PC board yourself, inspect it with an illuminated magnifier to ensure that there are no shorts or opens in the copper traces. Open any shorts with a hobby knife, and close any opens with thin solid wire and solder. Look closely, as solder balls can be teeny-weeny and still short closely spaced lands or pads.

- Check that the appropriate power supply voltage is present on every IC's power pins.

- Make sure your mechanical components are wired properly. An incorrectly wired jack on the input or output of your project can stop it dead in its tracks. The terminals on phone jacks with switch legs are not always arranged the same way, so you might have the input or output wire connected to the wrong terminal on the jack.

- If you severely distorted a plastic panel component during soldering or desoldering, it may now be kaput.

- Check that all ICs are plugged into their sockets properly. If a pin gets bent under an IC instead of plugged into its socket, it will definitely be a problem.

- Swap active components (op amps, logic chips, transistors) for known good ones (IC sockets to the rescue again). Cut the pins off of and throw out bad ICs (so you don't inadvertently use them again). Some people prefer to slingshot them as far as possible from the lab.

What to Look For (continued)

- The outputs of op amps used as comparators will always be in either high or low saturation. If the output of a comparator doesn't correspond with the measurements you observe on the inputs, you either have a bad IC or an issue with the components used to set the comparator's threshold or input level. A positive feedback resistor with an incorrectly low value will cause a comparator to *stick* in high or low saturation.

- The outputs of op amps used as amplifiers or filters should *never* be saturated high or low. If you come across one that is, make sure the associated resistor values are correct. If the feedback resistor's value is incorrectly high, it could cause the op amp's output to be saturated high or low. Knowledge of op amp theory (see Appendix A) will help you to determine if the output you are observing corresponds with the signals or voltage levels on the op amp's inputs. Consider the signals or levels observed on the inputs, along with the op amp's gain and biasing, and then determine if the output is appropriate. Look at the values of resistors associated with the op amp's circuitry to verify their values. If all component values and input signals are correct, replace the op amp.

- If using an NPN transistor as a switch to drive an LED, the transistor should allow current to flow from collector to emitter when the base is biased high. If the transistor doesn't respond to changes in base voltage (and subsequently current), and the resistors in the circuit are the correct values, replace the transistor.

- When using an N-Channel JFET in a high impedance source follower application (where its drain goes to V+ and its source goes to V- through a load resistor), you should see the voltage vary on the load resistor in the same manner as observed on the JFET's gate although the DC level may be different. If the voltage on the load resistor is not varying, make sure the resistors in the circuit are the correct values—and if they are, replace the JFET for a known good one.

- Op amps with gains higher than 1 can suffer from instability and oscillate at a high frequency (usually super-audio). Sometimes this oscillation will be heard as noise on the op amp's output, but in other cases you will need an oscilloscope to observe it. Using an oscilloscope, this high-frequency oscillation can be seen superimposed on the normal signal at the op amp's output. A small value capacitor (4.7pF to 100pF) placed across the op amp's feedback resistor may quell the oscillations.

- Leaky capacitors in timing applications will reveal themselves by discharging too quickly or never charging to full voltage when being charged by a low current. Tantalum capacitors are permanently damaged by being reverse-biased and become leaky very quickly, even with short exposure to incorrect polarity.

- Shorted capacitors will show the same DC level on both sides of the cap and will show up as shorted when tested with an ohmmeter.

- Low-quality potentiometers can suffer from poor conduction between the wiper and the resistive element, resulting in crackly operation in mixers or other odd behavior associated with *dead* spots in the pot's rotation.

- You will occasionally find switches that provide intermittent contact between terminals. Replace them.

- CMOS logic chip outputs should be at supply V+ or supply ground. If you observe an intermediate level on a CMOS chip's output, it could be that the circuit it is driving is too low in resistance (check values) or that the chip is defective and needs to be replaced.

- Anytime you want to see what's going on with an IC's pin while it is disconnected from the circuit, remember you can remove the chip, gently bend the lead of interest out, reinsert the chip, and probe the now isolated lead.

- LEDs are extremely reliable, so always check the driving circuitry first if you have an LED that won't come on.

Divide and Conquer

During troubleshooting of a complex circuit that contains many ICs and modules, it's often difficult to see the forest for the trees. In a case like that, it is necessary to divide and conquer. You may need to remove all of the chips from the PC board and then reinstall them one module at a time so you can test each module in isolation.

Isolate the power supply if you think it is the problem. Disconnect it from the circuit and see if it works under a reasonable load. Keeping some 1 to 10 watt ceramic resistors in low values (10 to 100 ohms) around is good for exercising a power supply.

With patience and determination (and perhaps a wee bit o' luck), you'll find and solve the problem that's keeping your project from coming to life. And never forget that a second set of eyes is always recommended. Cultivate relationships with relatives, teachers, professors, and friends who possess electronics knowledge so you'll have a resource to bounce ideas off of when you're in a tight spot.

Analog Synthesizer Building Blocks

3

In this chapter I'll cover a number of analog synthesizer fundamentals and then present the main building blocks or modules that provide the analog synthesizer with its incredible sound-shaping capability.

Three Configurations: Normalized, Modular, and Hybrid

Analog synthesizers generally fall into three categories: those with normalized switching schemes, those with modules and patch cords, and those that are a hybrid of the two (see Figure 3-1).

A synthesizer with a *normalized* switching arrangement lets the user connect the internal modules in a straightforward way using switches and potentiometers on the front panel. With this type, the user can only connect modules in the way the designer of the synth thought was most useful. That may be a slight limitation, but you can still do a ton of fantastic sonic work with one of these. Two of the most popular analog synthesizers of all time (the Mini-Moog and the ARP Odyssey) fall squarely into this category. The fact that these types are ready to go as soon as you turn them on and that they don't require patch cords is offset by the fact that they're generally a lot less flexible than their modular counterparts.

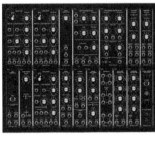

Figure 3-1. *Analog synthesizer modules vary in size, but are generally about two to four inches wide by eight or so inches tall. The faceplates bristle with cryptic legends, knobs, switches, buttons, and LEDs. Each one performs a particular function and has clearly marked inputs, outputs, and controls. The Mini-Moog (left) is an example of a normalized synth. The Synthesizers.com synthesizer (center) is a modular synth, and the ARP 2600 (right) is a hybrid of the two types.*

A completely *modular* synthesizer usually has a wooden case (or several) with tens (or even scores) of separate electronic modules in each one that the synthesist interconnects with patch cords. Patch cords are coax cables with phone plugs on both ends that fit into the jacks of the modules, permitting them to be interconnected. Some synthesizers use banana jacks instead of coax plugs to interconnect the modules. An experienced synthesist knows just how to connect the modules to produce whatever sound he is after.

37

The third type of synthesizer, the normalized–patch cord *hybrid* type, attempts to add the flexibility of patch cord interconnections to the normalized model. The famed ARP 2600 falls into this category. Although the front panel has a completely normalized switching scheme, allowing the synth to operate without patch cords, it also has jacks that allow the user to interconnect the modules in ways only possible with a completely modular synthesizer. It is generally a good idea to limit module interconnections so that outputs go to inputs. Doing otherwise can result in unexpected and possibly damaging results.

Voltage Control

Another important feature of analog synthesizers is that they are "voltage-controlled." The main voltage-controlled building blocks are the following:

Voltage-Controlled Oscillators (VCOs)
Varying voltage is used to control the frequency of a sound-producing oscillator.

Voltage-Controlled Filters (VCFs)
A voltage controls the cutoff frequency of a filter circuit.

Envelope Generator
A voltage may control the times for each of the attack, decay, sustain, and release phases.

Low-Frequency Oscillators (LFOs)
Varying voltage is used to control the frequency of a low-frequency modulation-producing oscillator.

Voltage-Controlled Amplifiers (VCAs)
A voltage controls the output amplitude of a signal.

A "one volt per octave" standard was developed by Robert Moog and was subsequently used by the majority of major analog synthesizer manufacturers. To illustrate what this means, consider this example. If you apply 1V to the control voltage input of an oscillator and note its frequency

(we'll imagine 100 Hz for our example), when you increase the control voltage to 2V, the oscillator will put out a frequency precisely an octave higher (200 Hz in this case). If you raise the voltage to 3V, the oscillator will put out precisely 400 Hz, another frequency increase of one octave. The change in the VCO's frequency as its exponential control voltage is increased in 1V increments is illustrated in Figure 3-2.

Expo CV = 0V, VCO Frequency = 100 Hz.

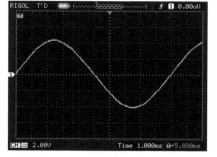

Expo CV = 1V, VCO Frequency = 200 Hz.

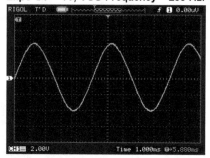

Expo CV = 2V, VCO Frequency = 400 Hz.

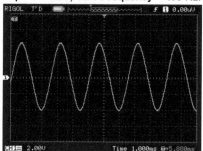

Figure 3-2. VCO frequency doubling as control voltage increases by 1V increments

In order to play notes that sound "in tune," the VCO's circuitry has to achieve a very precise conversion of control voltage to current. The ability of a VCO to "track" the control voltage accurately is extremely important.

Analog synthesizer VCOs are delicate pieces of circuitry that require temperature compensation and careful calibration to be used for musical purposes. The human ear can tell when something is even slightly out of tune. The "cent" system divides the distance between two adjacent semitones on the equally tempered scale into 100 divisions, and your ear can hear when your synth's VCO is only a few cents out of tune. VCFs that double as sine wave oscillators use similar circuitry to achieve cutoff frequency stability and precise volts per octave tracking. The ear's perception of loudness is not anywhere near as precise as its perception of frequency, so VCAs (the modules used for loudness shaping in a synthesizer) do not need to employ as precise a conversion mechanism as VCOs or VCFs.

The range of signal and control voltage amplitudes used within a synthesizer will vary depending on the manufacturer. They can range from +/–5V to as high as +/–15V. If you interface your synthesizer with another one, be sure not to exceed any control voltage limits, or you could damage one or both units.

The voltage used to control the modules of an analog synthesizer can come from several sources, but often a *keyboard controller* is used for musical applications. One of the outputs of the keyboard controller is a precise one volt per octave "control voltage" used to "play" VCOs designed to operate from the one volt per octave standard. We'll go further into the details of the keyboard controller below, but it will be instructive to briefly touch on two other signals supplied by the keyboard controller before we move on: *gates* and *triggers*.

Next let's explore the basic analog synthesizer modules and how each contributes to the amazing analog sound.

Figure 3-3 shows the symbols used in the analog synthesizer block diagrams that follow. Potentiometers and switches are typically used to set up a module's initial conditions. The module's operation is then dynamically changed by modulating voltages applied to its control voltage inputs.

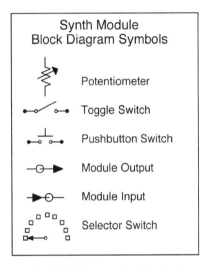

Figure 3-3. *The block diagrams in this chapter use these symbols to represent the controls found on each of the modules discussed.*

Gates and Triggers

Many analog synthesizer modules need to be activated with an electrical signal known as a *gate* (usually a voltage step from zero volts to some positive voltage) and/or a *trigger*, a few milliseconds-wide pulse that also typically goes from zero volts to some positive voltage. Figures 3-4 and 3-5 show typical gate and trigger signals, respectively. For instance, an envelope generator (which is often used to control a VCA to produce an amplitude envelope) needs to be activated when a key is pressed or some other controller is signaling that a note should sound.

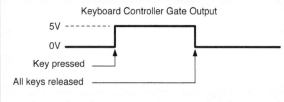

Figure 3-4. *Keyboard controller gate signal*

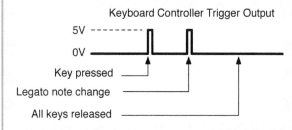

Figure 3-5. *Keyboard controller trigger signal*

In the case of the keyboard controller, when a person presses a key, the gate goes high. This gate output signal is connected via a patch cord to the gate input of the envelope generator, and when the gate goes high, the envelope generator begins its attack cycle . As long as the key is held down, the envelope generator continues to provide a voltage to the VCA, causing it to continue passing the signal applied to its input. When the key is released, the envelope generator enters its release or decay cycle, and the VCA's gain reduces to zero as the control voltage goes to zero at the rate set by the Decay or Release control of the envelope generator. This cycle is described in more detail later in Figure 3-15.

A trigger pulse a few milliseconds long is emitted from the keyboard controller any time a key is initially pressed and any time the note is changed (as in legato-style playing). A trigger pulse can also cause an envelope generator to spring into action, but with slightly different results that we'll detail below. I'll be referring to gates and triggers below, and now you'll know what I'm talking about.

Never let it be said that you won't come across some old piece of analog synth gear that uses negative voltage for gates and/or triggers or starts negative and goes positive (or weirder yet...*vice versa*). Always use an oscilloscope or multimeter to measure the output of a gate or trigger on a piece of gear you are unfamiliar with before connecting your project to it, just to be sure it's not one of these more *unusual* units.

The Voltage-Controlled Oscillator (VCO)

Figure 3-6 presents the block diagram of a voltage-controlled oscillator, a staple of all analog synthesizers. It is the main tone-producing module, and large analog synthesizers often include several of them. Sound is produced by vibrating the air around us, and sound waves propogate through the air as variations in air pressure. The shape of the vibrations makes a huge difference in the harmonic characteristics and timbre (tone color or tone quality) of the sound.

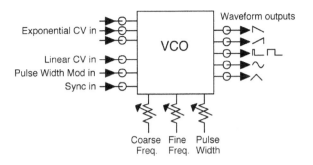

Figure 3-6. *VCO module block diagram*

Voltage-controlled oscillators generally produce several "waveforms," which are differently shaped impulses with which to vibrate the air. Typical waveforms are shown in Figure 3-7.

Each of these waveforms has a unique set of harmonics that give it a distinctive timbre. With experience you'll be able to identify each simply by hearing it.

Sawtooth

The sawtooth wave is a rather harsh sound and is often used in horn patches.

Ramp

The mirrored twin of the sawtooth, the ramp wave, sounds similar and is often at home in string patches.

Square

The square wave, with its infinite set of odd harmonics, is characteristically hollow-sounding. Some oscillators permit the square wave's duty cycle to be controlled and even modulated to produce the slow beating effect of two closely tuned oscillators. As you adjust the pulse width, you definitely hear when the duty cycle goes through the 50% point. Your eardrums just know when they're being vibrated back and forth evenly by the symmetric hollow-sounding square wave.

Rectangle

The rectangle wave is differentiated from the square wave by the fact that its duty cycle (high time versus low time) is *not* 50%. The more narrow the pulse width, the more nasal-sounding this waveform becomes, as more and more even harmonics are accentuated. The narrow rectangle/pulse wave always reminds me of the timbre of a harpsichord.

Sine

The sine wave is the "purest" waveform known to man and occurs frequently in nature. Any waveform can be built from sine waves occurring at the proper harmonic frequencies and amplitudes, and all waveforms have a fundamental frequency that is, you guessed it, a sine wave. As you can imagine just by looking at the rounded shape with no quick changes in direction, the perfect sine wave has no harmonics, and thus you only hear its fundamental frequency. However, since there are no *perfect* sine waves, suffice to say that sine waves have the fewest and lowest amplitude harmonics.

Triangle

The triangle wave is the sine wave's slightly dirty cousin in that its shape has far more harmonics than the sine, but it's a pretty mellow wave by itself.

If all the VCO did was produce a variety of waveforms, it wouldn't be that useful. One of the main features of a VCO is its response to a control voltage. We discussed one volt per octave control voltage previously, and VCOs must be carefully calibrated to achieve this goal.

Waveform Name	Wave Shape
Sawtooth	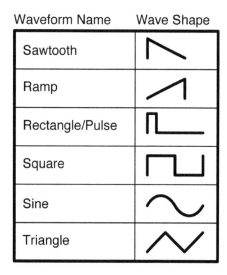
Ramp	
Rectangle/Pulse	
Square	
Sine	
Triangle	

Figure 3-7. Typical VCO waveforms

Once calibrated, a VCO can track one volt per octave over eight or more octaves. A VCO typically has potentiometers for both coarse and fine adjustments in frequency. The coarse control permits changing the frequency over the VCO's full range, while the fine control permits delicate tuning adjustments. The range of the VCO is typically the same as that of human hearing (20 Hz to 20 kHz), but many VCOs can go much lower in frequency and can be used as sources of low-frequency modulation.

> *Frequency is measured in cycles per second, or Hertz (after the German physicist Heinrich Hertz). It is the number of times per second that the sound source vibrates your eardrums back and forth. The abbreviation kHz stands for thousands of cycles per second, MHz stands for millions of cycles per second, and GHz stands for billions of cycles per second. Audible frequencies lie between 20 Hz and 20 kHz.*

A VCO typically has two types of control voltage (CV) inputs: *exponential* frequency response inputs and *linear* control voltage inputs. Whereas the exponential control voltage inputs change the VCO's frequency at a rate of one volt per octave, the linear control voltage input changes the frequency in evenly stepped increments. For example, if the VCO is oscillating at 100 Hz at 1V, an increase to 3V on the linear input would cause the frequency to rise to 300 Hz. A similar change on the exponential input would cause a doubling with each volt, with a resulting frequency of 400 Hz.

> *Often the linear control voltage input on a VCO is used for low-level modulation like vibrato.*

The voltage control inputs to a synthesizer module are often referred to as *modulation inputs*. Modules with dynamically changing outputs are typically used to change or "modulate" the frequency of the VCO. Another typical modulation input on a VCO is the *pulse width modulation* input. Usually there is a Pulse Width control potentiometer on the VCO with which you can set a square or rectangle pulse wave's output to a desired width (and resulting timbre). Having the ability to modulate the pulse width with a low-frequency sine wave (for instance) can make one VCO sound like two closely tuned VCOs beating against each other.

In Figure 3-8, you'll see the square wave voltage output of an LFO modulating the frequency of a VCO. The LFO's square wave is the blue trace, and the VCO's modulated frequency is the yellow trace. When the amplitude of the LFO's output is low, the VCO's frequency is correspondingly low; and when the amplitude of the LFO's output goes high, the VCO's frequency goes correspondingly high. This is an example of *voltage control* of the VCO's frequency.

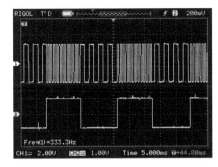

Figure 3-8. VCO frequency modulated by LFO square wave

In Figure 3-9, the VCO's frequency (yellow trace) is modulated by a ramp wave (blue trace). You can see how the VCO's frequency follows the amplitude of the ramp wave's voltage. As it gets higher the frequency increases and when it returns low the frequency follows. This is another example of *voltage control* of the VCO's frequency.

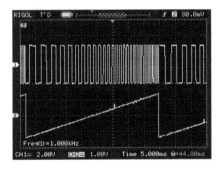

Figure 3-9. VCO frequency modulated by LFO ramp wave

Voltage-controlled oscillators come in two flavors: *ramp core* and *triangle core*. Even though VCOs can output several waveforms, the one that is produced in the bowels of the oscillator prior to being modified by wave shaping circuits is typically a ramp wave or a triangle wave. These two waveforms are able to be generated at precise frequencies from current sources and are great starting waveforms to put through wave shape modification circuitry to generate other waveforms.

As the name implies, a ramp core oscillator uses a precision circuit for generating a ramp wave

whose frequency is directly proportional to the current flowing in the VCO's linear voltage to exponential current converter. The triangle (or tri) core oscillator uses a precision circuit for generating a triangle wave whose frequency is, again, directly proportional to the current flowing in the VCO's linear voltage-to-exponential current converter. The VCO has circuitry to modify the shape of the core wave into the other waveforms shown in Figure 3-7.

Another interesting feature many VCOs possess is the ability to be synchronized (or *synced*) to another VCO. The output of one VCO (the sync source) is connected to the sync input of another VCO (the synced VCO), and the synced VCO's core oscillator is reset every time the sync source VCO finishes a cycle. The effect of this is that the synced VCO sounds like an emphasized set of harmonics, with its fundamental frequency set by the sync source VCO.

There are two types of sync: hard sync, in which the VCO's ramp core is reset to zero by the sync source; and soft sync, in which the synced VCO's triangle core oscillator's direction is reversed by the sync source. Hard sync is a much more in-your-face sound effect than soft sync. By sweeping the frequency of the synced oscillator, you get an unusual dynamic change in its timbre.

The Voltage-Controlled Filter (VCF)

The resonant voltage-controlled filter is one of my favorite modules because of its incredible sound-shaping capability. Figure 3-10 shows the block diagram for the versatile *state variable* type of VCF, which provides three response types simultaneously (low pass, band pass, and high pass).

The "wah-wah" sound that can be heard in everything from guitar foot pedals to the largest modular synthesizer is the result of a VCF. Voltage-controlled filters remove harmonics from the original signal, and because of this are often referred to as *subtractive synthesis* elements.

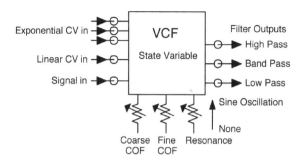

Figure 3-10. *VCF module block diagram*

This oscilloscope image in Figure 3-11 shows a square wave being put through a VCF. The VCF's cutoff frequency is high and the resonance is low, and thus the square wave passes through the filter virtually unchanged.

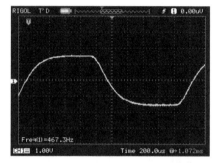

Figure 3-12. *VCO through VCF (lower cutoff frequency, low resonance)*

VCFs typically have adjustments to set the cutoff frequency (often with coarse and fine adjustments), and to adjust the amount of resonance, the filter adds to the signal being passed through them. They also have several control voltage inputs, both exponential (i.e., one volt per octave) and linear. Some filters include a convenient multichannel signal mixer as an integral part of the module.

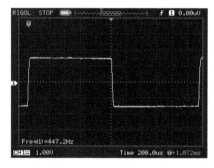

Figure 3-11. *VCO through VCF (high cutoff frequency, low resonance)*

In the oscilloscope image in Figure 3-12, the VCF's cutoff frequency has been turned down, which *subtracts* much of the higher harmonics from the square wave. If we continued to turn the cutoff frequency down, only the fundamental frequency of the square wave (appearing as a sine wave) would be passed. The filter's cutoff frequency could be adjusted even lower, completely blocking the square wave's fundamental frequency if we desired.

The VCF's exponential control voltage inputs that control the VCF's cutoff frequency follow the one volt per octave standard. When a VCO is put through a VCF, the VCF's cutoff frequency can track modulating voltages, just as the VCO does. This facilitates adjusting the VCO's frequency and the filter's cutoff frequency and resonance for interesting timbres, which can track across the entire range of a keyboard controller.

In addition to keyboard voltage modulation, the VCF can be modulated by envelope generators,

VCF Ringing

When the VCF's resonance is turned up, the filter emphasizes the center frequency of the signal being passed through it, resulting in "ringing" (see Figure 3-13). The ringing results from harmonic sinusoidal elements that are added to the sound passing through the filter. This ringing is what gives the resonant VCF its signature sound. The frequency of the sinusoidal ringing is the cutoff frequency in a low- or high-pass filter and the center frequency in a band-pass filter. The range of the VCF's cutoff frequency typically starts from well below audible and goes up to about 18 kHz or so.

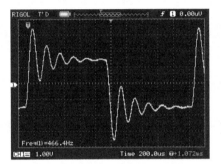

Figure 3-13. *VCO through VCF with high-resonance "ringing"*

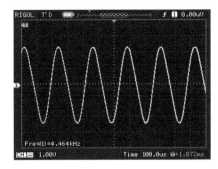

Figure 3-14. *A VCF with maximum resonance, resulting in oscillation*

The Voltage-Controlled Amplifier (VCA)

All sounds have amplitude envelopes. All sounds start, sustain, and end with a particular amplitude signature that helps us identify what they are. For example, the amplitude of a bell struck with a clapper starts high and then slowly decays over time until the bell's vibrations cease. A hammer striking an anvil has high initial amplitude that decays very rapidly. The amplitude of the sound of a cello or bass viol starts low and builds as the instrument's string begins to vibrate in response to the bow dragged slowly across it. The sound sustains as long as the bow continues to be dragged across the strings and can stop suddenly or slowly, depending on the actions of the musician.

Each of these phases of the amplitude envelope have names. The initial rise in amplitude is called the sound's *attack*. After the sound's *attack*, there is a *decay* phase in which the sound's amplitude falls to a lower level at which it *sustains* until it fades away over what we refer to as the *release* phase. Put together, these phases are referred to as the ADSR voltage envelope. You'll hear these terms again when we discuss envelope generators.

In Figure 3-15, we see clearly how the output of an ADSR envelope generator applied to the control voltage input of a VCA can create an amplitude envelope. The voltage envelope output by the ADSR is the blue trace, and the output of the

sample and hold modules, and LFOs: in short, any source of modulating voltage. The envelope generator is a popular modulator for the VCF and allows the synthesist to change the harmonic content of the sound dynamically in response to gates and triggers from a keyboard, a sequencer, and/or other types of voltage controllers.

On many VCFs, the output is a remarkably pure sine wave when the resonance is turned all the way up (see Figure 3-14). This allows the filter to double as a voltage-controlled sine wave oscillator. Under these conditions, the VCF's output can be used as a signal or modulation source.

VCA is the yellow trace. As a gate is applied and maintained, the envelope starts with the attack phase, followed by the decay phase, which falls into the sustain phase (the constant amplitude phase). Following removal of the gate signal, the release phase is entered, and the VCA's amplitude falls to zero along with the output voltage of the ADSR envelope generator.

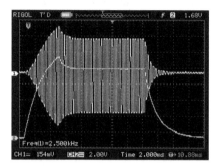

Figure 3-15. *ADSR modulating a VCA*

The VCA is the module that allows us to control the amplitude of the synthesizer's signal sources by means of a modulating voltage. The modulation source for the VCA is often an envelope generator (discussed in "The Envelope Generator (EG)" on page 48), a low-frequency oscillator, or a VCO. Often it is a mixture of many modulators (mixed using the DC mixer discussed in "Audio and DC Signal Mixers" on page 53).

VCA's typically have one or more signal inputs, an output, and several control voltage inputs. The VCA may include an Initial Amplitude control to allow the VCA to pass a set amount of the signal applied to its input. Applying a control voltage subsequently increases or decreases the amount of signal passing through the VCA. A Control Voltage Offset control is a useful feature that facilitates modulation from a variety of sources, some of which may be bipolar (oscillating about ground) in nature and others unipolar (biased either positive or negative). We'll talk more about *bias* in the sidebar "Biasing the Op Amp's Output" on page 128 in Appendix A.

VCAs can have both exponential and linear control voltages. Exponential VCA response sounds more organic than linear because humans perceive sound amplitude that way naturally. The block diagram of the VCA is presented in Figure 3-16.

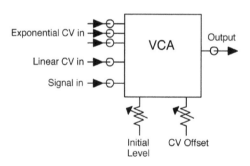

Figure 3-16. *VCA module block diagram*

The Low-Frequency Oscillator (LFO)

An LFO is like a VCO in many ways, except that it's in really slow motion. Whereas the typical VCO can go as low as a couple of cycles per second, the LFO's cycle time can be as long as several minutes. LFO oscillation can range from millihertz to audible frequencies (frequencies above 20 Hz).

LFOs typically have one or more frequency controls (coarse or coarse/fine) and can have several linear and/or exponential control voltage inputs. They typically produce all of the waveforms that a VCO can, including the variable-width-pulse waveform.

LFOs often serve as a modulator, producing vibrato in a VCO by applying the LFO's sine output (adjusted to about 10 or 12 Hz) to the VCO's linear control voltage input. LFOs can be used to modulate any voltage-controllable module. Some LFOs are capable of being reset with a gate or trigger pulse to sync the output with other events or modules. The block diagram is the same as that for the VCO in Figure 3-6.

The Keyboard Controller

In order to "play" the synthesizer like a keyboard instrument, you need a keyboard voltage controller. This controller has an organ-style keyboard with precision circuitry to cause the controller to output exactly one volt per octave. Keyboard controllers come in varying sizes: two octave, three octave, four octave, and all the way up to 88 keys, just like a full-size piano keyboard. A block diagram of a keyboard voltage controller is shown in Figure 3-17.

All key presses result in a control voltage output that is proportional to where on the keyboard the key resides. For example, say that pressing the lowest C on the keyboard resulted in an output voltage of zero volts, and pressing the C above that results in one volt. What about all the keys in between those two notes? Each key will output 1/12 of a volt (or 83.333mV) more than the next lower key. Table 3-1 shows one octave of control voltages.

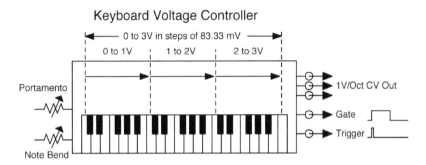

Figure 3-17. *Keyboard voltage controller block diagram*

Table 3-1. *One octave of control voltages (all units in mV)*

C	C#	D	D#	E	F	F#
0	83.333	166.666	249.999	333.332	416.665	499.998

G	G#	A	A#	B	C	
583.331	666.664	749.997	833.33	916.663	1V	

Keyboard controllers often include modulation wheels or joystick controls that allow a musician to augment the keyboard controller's output voltage with sine wave (or other waveform) modulation, as well as adding to or subtracting from the control voltage to produce note-bending effects.

Many synthesizer players feel that a keyboard is not expressive enough for the many sounds an analog synthesizer can produce. Ribbon controllers, light controllers, voltage sequencers, and knob twiddling are all viable alternatives to using a keyboard controller for producing electronic sound or musical compositions.

Mono Keyboard Controllers

Many classic keyboard controllers are *mono*, meaning they only allow one note to played at a time (no chords). Mono keyboard controllers usually have *low note priority*, which means that if more than one key is depressed at the same time, the voltage corresponding to the lowest key down will be emitted by the controller.

Some keyboard controllers allow playing two or more notes simultaneously and provide control voltage outputs that correspond to the keys being pressed. When used with a modular synthesizer, this feature permits controlling several oscillators, with the various control voltage outputs allowing the synthesist to produce chords.

Keyboard controllers that can output two separate control voltages can be designed with purely analog circuitry. Enabling a keyboard controller to output several control voltages, each corresponding to the nth note being pressed, typically requires a good deal of digital circuitry or a microprocessor and a scanning matrix style of keyboard.

A portamento or glide control is another feature typically present on keyboard controllers. When the portamento control is off, the keyboard outputs discrete voltages as the various keys are depressed. When the portamento control is advanced, the keyboard controller's output voltage slews from one level to the next as the keys are pressed. An excellent example of keyboard controller portamento is heard in the famous Moog solo performed by Keith Emerson in the Emerson, Lake, and Palmer song "Lucky Man."

If you were controlling a VCO with the keyboard controller and had the portamento control on, you would hear the notes slide from one to the other instead of changing pitch immediately. The more you advance the portamento control, the longer the slew between changing output voltages takes. Some keyboard controllers permit selection of either RC voltage slewing or linear voltage slewing. With RC slewing, the voltage changes respond in the same manner as a ca-

pacitor being charged or discharged through a resistor, whereas linear slewing (as its name implies), results in linear slewing from note to note.

Keyboard controllers also output two important control signals, known as gates and triggers, as discussed in "Gates and Triggers" on page 40.

The Envelope Generator (EG)

The discussion of the VCA touched on the subject of amplitude envelopes. The envelope generator is the module used to produce the actual voltage levels that correspond to the various amplitude envelope phases. The two most common types of envelope generators found in analog synthesizers are the attack release envelope generator (AREG) and the attack decay sustain release envelope generator (ADSREG). In Figure 3-18, you'll see a block diagram of the ADSR envelope generator.

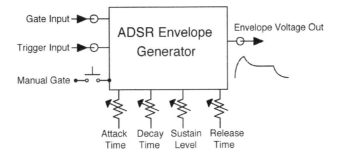

Figure 3-18. *ADSR envelope generator module block diagram*

AR Envelope Generator

The AREG has two phases: the attack phase and the release phase. Correspondingly, the AREG has Attack and Release Time controls that adjust the time the AREG takes to complete the attack and release phases. When adjusted to the lowest setting, the attack or release times are as short as possible (typically a few milliseconds). When adjusted to the fully on setting, the attack or release times are as long as possible (typically 20 seconds or more). The controls can be independently adjusted.

The AREG can be activated by either a gate or a trigger. When it is gated (i.e., a high gate signal is applied to its gate input), the AREG enters the attack phase, and its output voltage rises to its maximum (usually between 5V to 10V) at the rate set by the Attack Time control. The output of the AREG stays at that level until the user releases all keys, at which time the output of the AREG decays to its minimum output voltage at the rate set by the Release Time control. If the gate is removed prior to the AREG attaining its maximum output level during the attack phase, the module enters the release phase. The output voltage decays from whatever level it was at when the gate was removed back to its minimum level at the rate set by the Release Time control.

When the AREG is triggered (instead of gated) it starts and completes its attack cycle and then enters the release phase. Once in the release phase, the AREG can be retriggered to enter the attack phase and begin to climb to its maximum output level, at which time it enters the release phase and decays to minimum output.

ADSR Envelope Generator

The ADSREG can produce more interesting envelopes than the simpler AREG. Its ability to control the decay and sustain phases of its envelope provides the synthesist with more flexibility. The ADSREG responds to gates and triggers, so it can be gated, triggered, or gated and triggered simultaneously. We'll discuss the ADSREG's response in these three scenarios.

When the ADSREG is gated, it enters the attack phase, and the output of the module begins to climb toward its maximum output voltage at the rate set by the Attack Time control. If the gate is still applied at the completion of the attack phase, the ADSREG will enter the decay phase, and the module's output voltage will begin to decay to the sustain level at the rate set by the Decay control. If the gate continues to be applied throughout the completion of the decay phase, the ADSREG enters the sustain phase. The sustain phase is a user-settable DC level whose range is adjustable between the ADSREG's minimum and maximum output voltages (Figure 3-19 shows the ADSREG's output as a gate is applied throughut its attack, decay, and sustain phases). If the gate is removed during any of these phases (attack, decay, or sustain), the module enters the release phase, immediately skipping any phases between. Figure 3-20 shows the ADSREG's output when the gate is removed during the decay phase, and Figure 3-21 shows its output when the gate is removed during the attack phase. During the release phase, the module's output falls from the current voltage level to the ADSREG minimum output level at a rate set by the Release Time control.

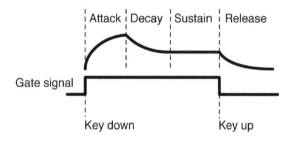

Figure 3-19. *ADSREG with gate applied throughout the attack, decay, and sustain phases*

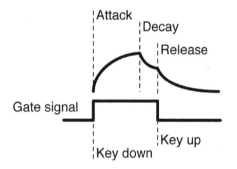

Figure 3-20. *ADSREG with gate removed during decay phase*

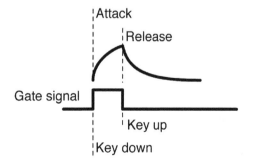

Figure 3-21. *ADSREG with gate removed during attack phase*

If the gate input is brought to the inactive state (low) and then brought to the active state (high), the ADSREG will reenter the attack phase, starting at the voltage level that the module's output had fallen to during the previous release phase.

Let's now consider applying only a trigger pulse to the ADSREG (instead of a gate). If the ADSREG is triggered, the ADSREG enters the attack phase and its output voltage rises to its maximum. As soon as maximum output level is reached, the module enters the release phase and begins to decay to the ADSREG minimum output level at a rate set by the Release Time control. If retriggered during the release phase, the module enters the attack phase and begins to rise toward its maximum output voltage again (from the level fallen to in the last release phase), at which time it immediately enters the release phase.

The ADSREG can be gated and triggered at the same time. When the ADSREG is triggered and gated at the same time, it responds just as it does when gated, with the only difference being that the trigger input can reset the ADSREG to the attack phase if the ADSREG is in any other phase (decay, sustain, or release). This is so that the module can be both gated and triggered with a keyboard voltage controller that emits triggers when the synthesist plays in a legato fashion (changing notes but never releasing all keys). When the synthesist changes notes but never releases all keys, a trigger pulse is emitted for each note change. Those triggers allow a perhaps percussive amplitude envelope to be emitted for each note change. Without this feature, the ADSREG would not enter the attack phase on note changes without all keys being released (see Figure 3-22).

ADSREGs have potentiometers for adjusting the time (or DC level in the case of the sustain control) for each supported *phase* (AR or ADSR). Some envelope generators have circuitry built-in to gate them at regular intervals (repeat gate), with the rate set by an additional *repeat rate* pot; or they have a *repeat* switch, which causes the envelope to automatically repeat at the end of the release cycle. Also normally present is a push button to apply a gate signal to the envelope generator manually.

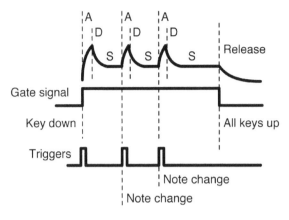

Figure 3-22. ADSREG with gate and triggers applied simultaneously

The White Noise Generator

The white noise generator is the module that produces the "shhhhhhhh" sound often used to produce wind, rain, thunder, drums, gongs, and so on. White noise is made up of all audible frequencies in an equal power distribution, similar to how white light is made up of all visible colors in an equal distribution. White noise can be converted to "pink" noise through specialized low-pass filtering. Raw pink noise generally sounds more thundering than raw white noise. Often, noise generators will supply outputs for both types.

Figure 3-23 shows the block diagram of the white (and often pink) noise generator.

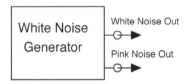

Figure 3-23. White noise generator block diagram

Making White Noise

White noise is most often generated by reverse-biasing the emitter-base (EB) junction of an NPN transistor with enough voltage to exceed its breakdown voltage. The noise produced at the EB junction is then amplified with high gain in order for the amplitude of the white noise to be high enough to use as a signal or modulation source in an analog synthesizer.

White noise can also be generated digitally with shift registers and XOR logic gates, but the technique suffers from the fact that most digital white noise generators have a "period" over which they repeat. If the "period" is too short, the noise will have an unfortunate cyclic character that you can easily hear. The reverse-biased EB junction technique produces a never-repeating series of random voltages and thus no cyclic artifact, as shown in Figure 3-24.

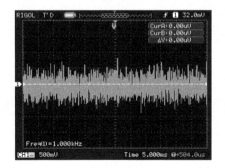

Figure 3-24. White noise waveform

The Sample and Hold (S&H)

The sample and hold is an often misunderstood module. It is not something that you normally use to put a signal "through" to produce an effect. Usually the output is used to modulate other voltage-controlled modules. The S&H has a Sample Rate control, which is the rate at which it takes a millisecond or so duration snapshot of the voltage level on its input.

The way it works is this: an analog switch of some type (CMOS analog switch, MOSFET, JFET) is used to route the voltage level at the input onto a capacitor (the voltage level storage medium) at regular intervals (the sample rate). The sampled level is held until the next sample is taken (thus, sample and hold). Figure 3-25 presents the sample and hold block diagram.

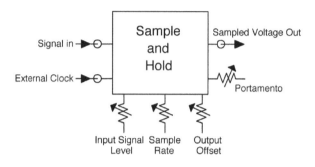

Figure 3-25. *Sample and hold module block diagram*

The input to the sample and hold can be the output of any other module: VCO, LFO, noise generator, ADSREG, etc. Sample rate is adjustable from one sample every few seconds to hundreds of samples per second. The effect of the S&H is to turn the waveform applied to its input into a stepped representation of that waveform. In Figure 3-26, the blue trace is the triangle wave output of an LFO, and the yellow trace is the output of the sample and hold, which is taking periodic samples of the LFO's output voltage and holding them until the next sample is acquired.

If we applied the LFO's triangle wave directly to the control voltage input of a VCO, the VCO's frequency would rise and fall in a continuous siren-like manner. However, as you may imagine from looking at the yellow trace in the scope photo of Figure 3-26, the sample and hold module's output applied to a VCO's control voltage input results in a series of discrete rising and falling steps in pitch.

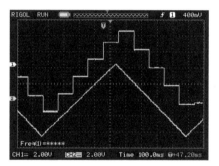

Figure 3-26. *LFO triangle wave being sampled and held*

In Figure 3-27, we see the output voltage of the sample and hold (yellow trace) that is sampling the ramp wave output of an LFO (blue trace).

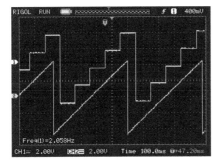

Figure 3-27. *LFO ramp wave being sampled and held*

Controls for adjusting input and output levels as well as a voltage offset are often provided on this module. When the rate of the signal applied to the input of the sample and hold is higher than the sample rate (for example, a VCO used for S&H input), interesting beating effects can be achieved. Another interesting use of the sample and hold is to apply white noise to its input. The instantaneous voltage level of white noise taken at intervals is random, so if you apply the output of the S&H that is sampling white noise to a VCO, you will hear a never-ending series of random tones. Figure 3-28 shows the random levels of output voltage put out by the sample and hold when white noise is applied to the input.

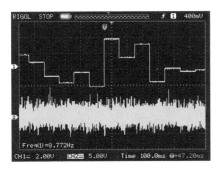

Figure 3-28. *White noise being sampled and held*

The S&H also has multiple control voltage outputs, glide adjustment (sliding between notes effect), and gate and trigger outputs that go high synchronously with each new sampled output voltage. Sample and hold modules can also be clocked by an external clock source for more flexibility.

Audio and DC Signal Mixers

Mixers are the unsung workhorses of the analog synthesizer. Analog synths often have several audio signal sources: multiple VCOs, a noise generator, and possibly an external signal input. The signal mixer allows the synthesist to mix the outputs of these sources prior to further signal processing, such as filtering or amplitude modulation. This is generally a simple mono mixer with level controls and perhaps mute buttons or switches for each input.

Although the synthesizer's signal outputs can be connected directly to an amplifier or mixing board, a large modular synthesizer often includes an integral stereo panning mixer with effects loops, equalization controls, stereo line outputs, and a stereo headphone jack. The outputs of this mixer can go straight to a stereo power amplifier or to the inputs of an audio mixing board.

The synthesizer's DC mixer is for combining the outputs of modulators whose levels change dynamically over relatively long periods of time (long relative to the periods of audio signals). Audio mixers typically block DC signals and offsets using capacitive coupling on the input and output, but DC mixers are designed without capacitive coupling and permit mixing of DC modulation signals such as multiple LFOs, ADSREGs, keyboard controller voltage, etc. A DC mixer will contain level controls for each input and may also include a DC bias control for adding a DC offset to the resultant mix (see the sidebar "Biasing the Op Amp's Output" on page 128 in Appendix A for more information). Additionally, the DC mixer may allow any or all of the inputs to be inverted (phase shifted by 180 degrees).

AC Versus DC Coupling

When an input is AC coupled (i.e., coupled through a capacitor), any DC offset the signal may have is blocked, since capacitors do not pass DC voltage. We illustrate this in **Figure 3-29**. *When a voltage is applied to an AC-coupled input, only the AC component of the signal is passed through the capacitor— and thus the voltage on the resistor, if observed on an oscilloscope, would be oscillating about ground (not about 2V). When the input is directly coupled (DC coupling), the 2V offset remains, as well as any AC signal riding on the DC offset.*

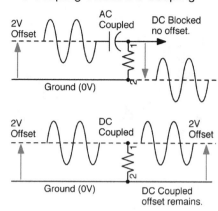

Figure 3-29. *AC versus DC coupling*

The block diagrams in Figure 3-30 show the differences between the audio signal mixer with its AC-coupled inputs and output versus the DC modulation mixer, which allows mixing of slow changing DC levels of several modulation signals.

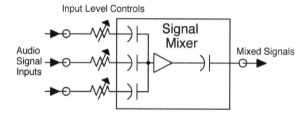

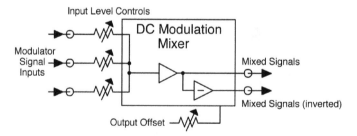

Figure 3-30. *Signal and DC modulation mixers*

The Ring Modulator

The ring modulator produces very interesting sonic effects by amplitude modulating or by multiplying one input signal by another. The multi-plying or modulating signal can be an adjustable frequency sine wave, the input signal itself, or another input signal. Although modern *ring* modulators often do not contain the *diode ring* of the early transformer-based circuits, the name has stuck. The block diagram of the ring modulator is presented in Figure 3-31.

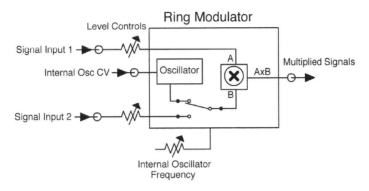

Figure 3-31. Ring modulation module block diagram

Two of the most popular methods of producing ring modulation are amplitude modulation and signal multiplication. The first is similar in effect to routing one signal through a VCA (for example, a 1000 Hz square wave) and then modulating the VCA's control voltage with another signal (say a 500 Hz sine wave). The output of the ring modulator would be a 1000 Hz square wave, with amplitude modulation at 500 Hz. The deeper the amplitude modulation, the more pronounced the effect becomes.

Figure 3-32 shows a 400 Hz sine wave and a 5 kHz square wave and the result of multiplying the signals together via a ring modulator (in this case the MFOS Sonic Multiplier). The resulting signal's timbre is very different from either of the original waveforms.

400Hz. Sine and 5KHz Square

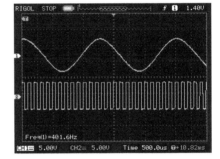

After multiplication via a Ring Modulator

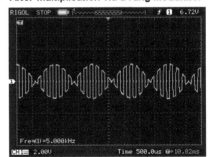

Figure 3-32. Sine and square wave through a ring modulator

By adjusting the modulating (or multiplying) signal's frequency, you can discover many interesting bell-like timbres. When the modulating (or multiplying) signal's frequency is modulated with an LFO, the effect is unique and must be heard to be appreciated.

The other method of producing the ring modulation effect is to use an analog multiplier chip (the AD633 from Analog Devices, for example) to multiply the two signals together. This produces a very similar sonic effect to the amplitude modulation method. Both methods result in the output containing the sum and difference of the frequencies contained in both signals, which produces the bell-like tone. Modulating one VCO's control voltage input with the output of another VCO can produce very similar-sounding timbres (in case you don't have a ring modulator yet). By experimenting with different waveforms and frequencies for both the carrier (input signal) and the modulator (modulating signal), you will discover a wide range of interesting timbres.

The Voltage Sequencer

Another extremely cool module is the voltage sequencer (usually just called a *sequencer*). A sequencer is a module that sequentially steps through a series of user-adjustable voltages at a rate set by the user. So, for example, if the sequencer was being used to control a VCO, the VCO would repeatedly output a series of frequencies corresponding to the voltage settings output by the sequencer. The voltage output of the sequencer can be used to modulate any voltage-controllable module. Sequencers typically have eight or more *channels*, with each channel (or step) having an active channel (or step) indicator LED, coarse and fine control voltage adjustment pots, a gate on/off switch, and on some sequencers a time per step adjustment.

Sequencers generally have two modes. One is the manual mode, in which the user uses a push button on the sequencer's control panel to *step* the sequencer forward or backward through the various channels (or steps) to adjust the coarse and fine voltage level for each step. The other is the run mode, in which the sequencer automatically steps through the series of voltages at a rate determined by the user. The sequencer's *clock rate* (rate at which the sequencer steps through the series) is adjustable by means of a front panel pot (or an external clock if the sequencer supports it). A block diagram of the voltage sequencer is presented in Figure 3-33.

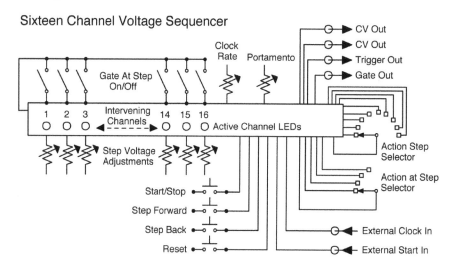

Figure 3-33. *Voltage sequencer block diagram*

The sequencer allows selection of the number of steps in the series (usually from two up to the maximum number of *channels*) and the sequence type. The sequence type can vary based on user setup and can sequence start to end and stop or sequence repeatedly: start to end; start to end, then reverse direction to start and continue; or randomly (any channel has an equal chance to be active). Some sequencers provide additional sequence types. For example, a 16-step sequencer may have a mode in which it allows 8 steps with two separate output voltages per step, channels 1 and 8 being active together, followed by 2 and 9, 3 and 10, etc.

Additionally, there are gate and trigger outputs that go high synchronous with the output voltage changes that can be used to control envelope generators or other modules. Sequencers often provide circuitry to allow external clock signals to start, stop, and/or advance the sequence.

Portamento control, another common feature, enables the gliding effect between voltage steps.

Sequencers can be set up so that the gate from a keyboard controller starts the sequence via its external start input. If the sequencer's mode is set to sequence and then stop on the last step, an arpeggio effect is produced every time a key is pressed. Needless to say, sequencers are a blast to both build and play with. The MFOS 16-Step Sequencer in Figure 3-34 was built by synth-DIYer Girts Ozolins, of Latvia. Girts, a seasoned DIYer, made his own professional-looking panel and added a MIDI clock interface to the sequencer.

Figure 3-34. *MFOS 16-Step Sequencer*

Putting It All Together

The typical normalized analog synthesizer has one or more oscillators and a white noise generator to produce signals. Those signals are routed via level pots into an audio mixer, whose output is normally fed to the VCF. The output of the VCF is routed to the VCA, whose output is routed to the synthesizer's output. Several modulation possibilities are provided by the modulation signal routing.

A fully featured synthesizer will have two sets of modulation sources (ADSREG and LFO), one for

the VCF and one for the VCA. The modulation sources are also routed to the VCOs. Modulation level pots are provided for each module to adjust from zero to full modulation from any source.

The keyboard controller applies one volt per octave control voltage to each VCO and to the VCF. We mentioned above that the resonant VCF, when adjusted to maximum resonance, can double as a one-volt-per-octave tracking sine wave oscillator. We have shown the same set of ADSR and LFO modulators as providing modulation for both VCOs, but a more flexible scheme would be to have switches to permit either oscillator to be modulated by either set of modulators.

Another desirable feature is to be able to route one of the VCO outputs to become a modulator for the VCA in order to achieve ring modulation effects. Permitting one of the VCOs to modulate at the frequency of the other one is another very-nice-to-have feature. I have not discussed every control shown in the typical normalized analog synthesizer architecture in Figure 3-35 due to space considerations. This rather simple configuration, even though it is normalized, can easily make an infinite set of unique sounds.

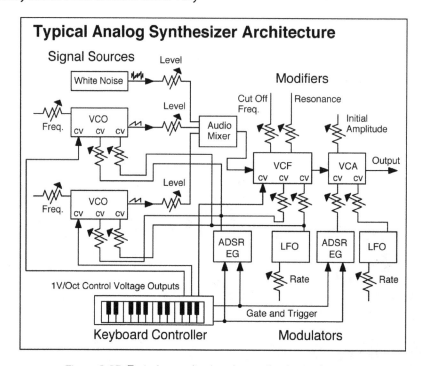

Figure 3-35. *Typical normalized analog synthesizer architecture*

Note that the setup in **Figure 3-35** *does not show the sample and hold, ring modulator, or additional audio or DC mixers in the diagram because they are not typically found in most normalized synths and fall more into the realm of the modular synthesizer.*

Remember that the modular analog synthesizer can have multiple VCOs, VCFs, VCAs, modulators, noise generators, ring modulators, waveform modifiers—in short, as many modules as you want to put into your synth cabinets. Some modular synths fill an entire room, and not with large 1950s tube technology, but with modern transistor- and IC-based circuitry. Search for information on the "TONTO Synthesizer" if you want to see a jaw-droppingly huge modular synthesizer.

Make the Noise Toaster Analog Sound Synthesizer

4.

In this chapter we're going to get down to business and build a small analog synthesizer called the *Music From Outer Space Noise Toaster*. For many of you, this may be your first project, so I'll include many details that will become second nature to you as you gain knowledge of and experience with synth-DIY. I know for a fact that veterans (*including me*) find this project fun to build as well. For many of us, it's a reminder of when we were just getting started and life was simpler.

Before we get building, let's quickly review the Noise Toaster block diagram and learn a bit about its normalized switching scheme.

Noise Toaster Block Diagram

The Noise Toaster (Figure 4-1) has seven main modular components:

- White noise generator
- Voltage-controlled oscillator (VCO)
- Voltage-controlled low-pass filter (VCF)
- Voltage-controlled amplifier (VCA)
- Low-frequency oscillator (LFO)
- Attack release envelope generator (AREG)
- One watt amplifier

The white noise generator's output can be routed to the input of the VCF via the White Noise Input Select switch. Sounds such as wind, explosions, surf, and other white noise–based sounds can be created by the unit thanks to this module. A block diagram, highlighting the main modules of the Noise Toaster, is shown in Figure 4-2.

The VCO produces ramp and square wave outputs, and either waveform (or none) can be routed to the VCF via the Input Select switches. The VCO's frequency is set by the Frequency control and ranges from a few clicks to 10 kHz or more. The LFO Mod Depth control adjusts how much

the LFO modulates the VCO's frequency. The AR Mod Depth control adjusts how much the AREG modulates the VCO's frequency when the AR Mod switch is turned on. Finally the Sync switch, when on, allows the LFO to reset the VCO's ramp core, producing unusual timbral effects. The Sync effect is most pronounced when the LFO is near the top of its adjustment range and the VCO's frequency is swept up and down.

Figure 4-1. *The MFOS Noise Toaster*

The VCF accepts the output of the VCO and white noise generator as inputs. The VCF's Cutoff Frequency control adjusts the top end of the filter's response. As Cutoff Frequency is turned up, more and more of the input signal's higher harmonics are passed through the filter. The Resonance control adjusts the amount of ringing the filter produces. When Resonance is turned up, the VCF adds harmonic character to the signals being passed. The VCF's cutoff frequency can be modulated by either the AREG or the LFO (or neither), depending on the setting of the Mod Source switch and the Mod Depth control.

The unit's simple VCA is always modulated by the AREG but can be bypassed with the Bypass switch. When not bypassed, the VCA's amplitude is controlled by the AREG voltage envelope.

The LFO produces three selectable output waveforms: Square, Differentiated Square, and Integrated Square. The Frequency control sets the LFO's repetition rate, which ranges from less than 1 Hz to over 200 Hz. The output of the LFO can be switched to modulate the VCO and/or the VCF.

The AREG produces a voltage envelope that is used to modulate the VCF and the VCO. The Attack control adjusts how fast the voltage envelope rises and the Release control adjusts how fast the voltage envelope fades away. Attack and decay time adjustments range from a few milliseconds to several seconds. The AREG can be gated manually when the Repeat/Manual switch is in the Manual Gate position and the Manual Gate button is depressed. Holding down the Manual Gate button will cause the AREG to rise to its maximum voltage and stay there until the button is released. When set to Repeat (Repeat/Manual switch in Repeat position), the AREG behaves like an LFO whose output waveform shape and frequency can be adjusted with the Attack and Release controls.

Finally, the unit's battery-powered one watt amplifier adds portability to the unit, allowing you to play it anywhere. When a cable is plugged into the unit's output jack, the internal amplifier and speaker are bypassed. The volume control adjusts the output level whether using the internal amplifier and speaker or connecting the unit to an amplifier or mixing board.

The whole thing is powered by one 9V battery, but you can also use a clean 9V power supply if you choose to. Now you have an overview of what makes up the Noise Toaster. We'll discuss the Noise Toaster schematics and theory of operation in much greater detail in Chapter 6. I decided to put the Noise Toaster construction chapter first so you can dive right in and get *making*!

The finished project makes a great portable sound generator and an excellent gift for that imaginative young person in your life. You'll never be unprepared for an impromptu sound battle or performance art opportunity again.

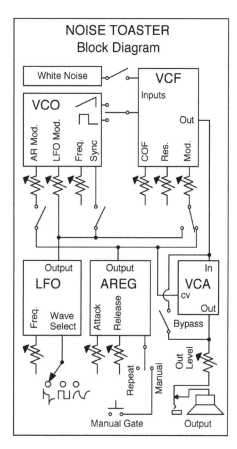

Figure 4-2. *Noise Toaster block diagram*

Building the Noise Toaster

I can't emphasize enough how much reading through this chapter and visualizing yourself succeeding with the project before picking up a tool or component will help you. You will avoid putting yourself into one of those glue-on-your-fingers, not-enough-hands moments_ where you say..."Oh drat, I wish I had known this before now."

The first thing you'll need to do is gather the parts by reviewing the project part list (also known as a BOM or bill of materials) and purchasing them from your favorite electronic component distributor. I'll include some suggestions regarding obtaining parts online from sources I have used for years with excellent success.

Noise Toaster PCBs, component kits, and faceplates can be made from the images in this section or purchased from the **Music From Outer Space website** *(http://www.musicfromouterspace.com).*

All MFOS PCBs are professionally manufactured, double-sided glass epoxy material with plated through-holes, solder-coated donut pads, solder masking, and silk-screened component legends. The PC patterns for the top copper layer, bottom copper layer, and silk-screened legend are provided in this chapter to enable you to make your own PCB if you wish. I do not cover PCB fabrication, so you need to know how to make double-sided PCBs if you decide to make your own.

I suggest you make a simple case out of wood as shown in Figure 4-1. If you don't have the facilities to fabricate a case, an alternative would be to buy an aluminum chassis and construct the unit inside it. Mouser Electronics sells a 7" long by 7" wide by 2" deep chassis from manufacturers BUD and LMB that will house the project perfectly. You can purchase the chassis alone or buy the chassis and separate chassis bottom plate to completely enclose the project. Table 4-1 includes the Mouser Electronics part numbers for both chassis and cover plates.

Table 4-1. Chassis parts

Chassis	Mfgr's Part No.	Mouser Part No.
7" x 7" x 2" BUD Chassis	BUD #AC-405	Mouser #563-AC-405
7" x 7" BUD Chassis Bottom Plate	BUD #BPA-1952	Mouser #563-BPA-1592
7" x 7" x 2" LMB Chassis	LMB #772	Mouser #537-772
7" x 7" LMB Chassis Cap Cover	LMB #772C	Mouser #537-772C

If you fabricate your own case, I suggest you make the front panel from conductive material (aluminum or stainless steel). Since the output jack is mounted on the front panel, its ring (ground connection) is in contact with the front panel, providing it with a convenient grounding point. It is important to make sure that all of the potentiometer bodies are connected to the circuit's ground, and having the front panel made of metal and grounded by means of the output jack works perfectly.

I'll present a suggested front panel layout and accompanying wiring diagram; if you are confident in your electronic skills, use your imagination and design a front panel with your personal touch. If you opt to design your own panel, I suggest that you also make an accompanying wiring diagram to reference while constructing the unit. The suggested panel layout and accompanying wiring diagram presented covers both the intra-panel wiring and the PCB to panel wiring.

Last but not least, we'll cover testing the completed unit, along with some troubleshooting tips in case things don't go as planned.

Gathering the Parts

Before we begin to detail the operation of the Noise Toaster's circuitry, I want to cover some specs shared by all the parts. In the circuit descriptions that follow, instead of clogging up the text with repeated component specifications, e.g., R58 (1/4W 5% 100K resistor), I would like to state the following:

- *All* resistors are 1/4W 5% carbon film or carbon composition.

- *All* capacitors must have a voltage rating of at least 16V.

- *All* potentiometers are *linear taper*, with minimum power rating of 250mW.

- *All* diodes are 1N914 high-speed switching type or equivalent.

- *All* switches are mini-toggle type.

I will cite a component's value or specification in parentheses on its first reference within a section but thereafter by designator only. Switch type *SPDT* means *single pole, double throw*, type *SPDT C.O.* means *single pole, double throw - center off*, type *SPST* means *single pole, single throw*, and finally type *SPST N.O. PB* means *single pole, single throw - normally open - push button*.

I don't have a good electronics shop within a reasonable distance of my home, so I regularly buy my components online. I find buying electronic components online very convenient, and I can usually find anything I need for a reasonable price. However, if you happen to live in an area where electronic shops with good prices and wide selection abound, by all means give the business to the local shops.

However, if they don't have a part you need, then I suggest you give online component shopping a go. I think once you go through the process, you'll find that most places provide excellent prices and fast delivery. Here are some suggested online component distributors I have used for years:

- Jameco Electronics (*http://www.jameco.com/*) has been around for years serving hobbyists and professionals alike with a wide selection of components and reasonable pricing.

- All Electronics (*http://allelectronics.com/*) is another great place to buy from, and if you live in the Los Angeles area, they have a huge retail store that is a blast to browse in. They sell surplus and new components.

- Mouser Electronics (*http://www.mouser.com/*) has been around for years and will send you their free catalog, which is three or four inches thick. They have *everything*.

- DigiKey (*http://www.digikey.com/*) is another company that sells everything electronic.

Below you'll find the list of parts you'll need to purchase to construct the Noise Toaster.

The Bill of Materials (BOM)

All of the parts listed in Table 4-2 have the following requirements, as applicable:

- All capacitors should have a working voltage rating of at least 16V.

- Capacitors with lead spacing of .2" (5mm) fit nicely on the PCB.

- All integrated circuit packages must be DIP type.

- The power rating of potentiometers should be at least 250 mW.

- Control knobs must fit the potentiometer shafts.

Table 4-2. Noise Toaster bill of materials

Active Components

Qty	Description	Value/Note	Designators
2	LM324 Quad Low Power Op Amp	14 pin DIP	U1, U2
1	LM386N4 (or N3) Low Voltage Audio Power Amp	8 pin DIP	U3
1	2N3906 PNP Transistor	T092	Q10
4	2N5457 N Channel JFET	T092	Q1, Q7, Q8, Q9
5	2N3904 NPN Transistor	T092	Q2, Q3, Q4, Q5, Q6
3	1N914 High Speed Sw. Diode	DO-35	D1, D2, D3
1	Red LED	General Purpose	LED1
2	14 pin DIP IC Socket	High Reliability	
1	8 pin DIP IC Socket	High Reliability	

Resistors and Pots

Qty	Description	Value/Note	Designators
7	Linear Taper Potentiometer	100K	R1, R13, R20, R28, R29, R38, R66
3	Linear Taper Potentiometer	1M	R49, R51, R53
3	Resistor 1/4 Watt 5%	2M	R5, R16, R30
1	Resistor 1/4 Watt 5%	300K	R7
1	Resistor 1/4 Watt 5%	36K	R21
1	Resistor 1/4 Watt 5%	15K	R18
2	Resistor 1/4 Watt 5%	39K	R39, R40
3	Resistor 1/4 Watt 5%	3K	R48, R65, R70
5	Resistor 1/4 Watt 5%	4.7K	R22, R23, R37, R42, R54
2	Resistor 1/4 Watt 5%	4.7M	R33, R35
3	Resistor 1/4 Watt 5%	470K	R10, R19, R41
2	Resistor 1/4 Watt 5%	47K	R27, R36
1	Resistor 1/4 Watt 5%	75K	R2
1	Resistor 1/4 Watt 5%	820 ohm	R50
1	Resistor 1/4 Watt 5%	10 ohm	R69
12	Resistor 1/4 Watt 5%	100K	R14, R24, R25, R31, R44, R45, R47, R55, R58, R59, R60, R67
11	Resistor 1/4 Watt 5%	10K	R4, R6, R9, R12, R15, R17, R26, R32, R52, R61, R62
2	Resistor 1/4 Watt 5%	150K	R63, R64
1	Resistor 1/4 Watt 5%	3M	R34
2	Resistor 1/4 Watt 5%	1M	R8, R46
1	Resistor 1/4 Watt 5%	200 ohm	R57
2	Resistor 1/4 Watt 5%	20K	R68, R56
1	Resistor 1/4 Watt 5%	270K	R3

1	Resistor 1/4 Watt 5%	27K	R43
1	Resistor 1/4 Watt 5%	2K	R11

Capacitors

2	Capacitor Aluminum Bipolar (nonpolarized)	1µF	C17, C21
3	Capacitor Ceramic	.001µF	C2, C9, C10
1	Capacitor Ceramic	.047µF	C24
8	Capacitor Ceramic	.1µF	C3, C4, C5, C7, C8, C11, C12, C20
1	Capacitor Ceramic	100pF	C1
1	Capacitor Ceramic	.01µF	C6
3	Capacitor Electrolytic	10µF	C13, C14, C25
2	Capacitor Electrolytic	1µF	C16, C19
1	Capacitor Electrolytic	4.7µF	C18
1	Capacitor Electrolytic Radial Lead 0.2" Spacing	220µF	C22
3	Capacitor Electrolytic Radial Lead 0.2" Spacing	470µF	C15, C23, C26

Hardware and Other Components

1	Speaker 2 1/4"	8 ohm 1W	SPK1
2	Switch SPDT (Center Off) Mini Toggle	SPDT C.O.	S3, S6
2	Switch SPDT Mini Toggle	SPDT	S8, S9
6	Switch SPST Mini Toggle	SPST	S1, S2, S4, S5, S7, S11
1	Switch SPST N.O. Pushbutton	SPST	S10
1	Jack 1/4" with NC Switch to Tip		J1
1	9V Alkaline Battery		B1
1	9V Battery Snap Connector	Radio Shack Model: 270-325	
1	9V Battery Holder	Radio Shack Model: 270-326	
1	Project Case 7x7x2"	Mouser Part# 563-AC-405 7" x 7" x 2" BUD Chassis	
2	6-32 1 1/4" Machine Screws	6-32 Thread	
4	6-32 Machine Screw Nuts	6-32 Thread	
4	1/4" W x 3/8" L Nylon Spacers		1
25' Spool of Stranded Wire	22 or 24 AWG		10

Figure 4-3 shows a variety of capacitor types, all of which are found in the BOM.

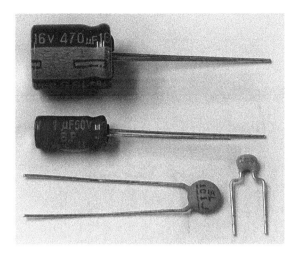

Figure 4-3. *Capacitor types. Note that C17 and C21 are nonpolarized capacitors, as shown at left. The polarized electrolytic capacitors have a polarity indicator (top), whereas the nonpolarized type are marked as B.P. (bipolar). Ceramic capacitors are also nonpolarized (bottom).*

The Noise Toaster Printed Circuit Board (PCB)

The Noise Toaster PCB is 5.2" long by 2.7" wide. The top layer's copper pattern (see Figure 4-4) and the silk-screened legend are as viewed from the top of the board. The bottom layer's copper pattern (see Figure 4-5) is as viewed from the bottom of the PCB. Be sure and expose the patterns in the correct orientation when making the board. Any text within the copper traces or silk legend (see Figure 4-6) will read properly when the pattern is correctly oriented.

Professionally manufactured, double-sided PCBs have plated through-holes, which are often used to route circuit connections between the top and bottom sides of the board. These top-to-bottom circuit connections are called *vias*. All pads on the board not used for component leads must be drilled and have a "Z" wire inserted and soldered. A "Z" wire is a piece of solid 22 or 24 gauge hook-up wire that is inserted through the pad's hole and then bent at a 90-degree angle on the top and bottom sides of the PCB. Trim any excess wire after bending, make sure the ends are not shorting to adjacent pads, and then solder the "Z" wire top and bottom.

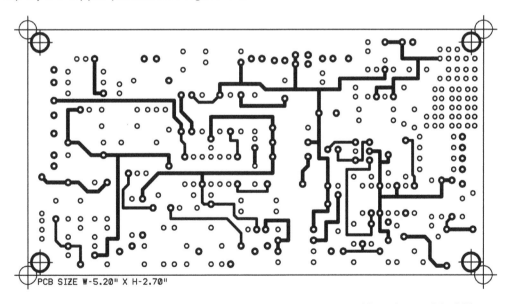

PCB SIZE W-5.20" X H-2.70"

Figure 4-4. *This is the Noise Toaster PCB's top copper layer as viewed from the top of the PCB*

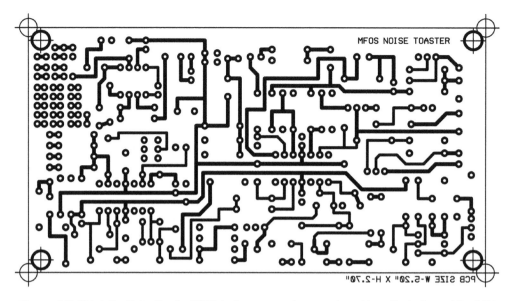

Figure 4-5. *This is the Noise Toaster PCB's bottom copper layer as viewed from the bottom of the PCB*

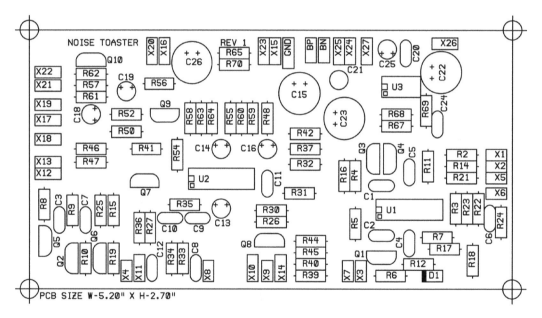

Figure 4-6. *This is the Noise Toaster PCB's top silk screen layer as viewed from the top of the PCB*

All circuit points with *Xn* designators throughout the schematics represent pads on the PCB, which are used to interconnect the populated PCB with the front panel controls. See the front panel wiring diagram in Figure 4-10 for full PCB to front panel wiring details.

Populating the Noise Toaster PCB

Now that you've obtained or fabricated the Noise Toaster PCB, it's time to populate it. Figure 4-7 is a view of the Noise Toaster PC board showing the component outlines overlaid with their values. It's just what you need for the PC board population phase of the project. I like to work from low components to taller components, so I start by installing the resistors and diodes onto the PCB. Insert a few components at a time, and then solder them. Prior to installation, bend each component's leads so that they align with the holes at the ends of its respective silk-screened legend. With practice, this will become easier to do. I like to hold the component between my thumb and forefinger and then with even pressure, bend both leads down with the thumb and forefinger of my other hand. There should be a small arc of lead next to each side of the component.

Once a component is installed, you should bend the leads so that the component will not fall out of the board when you turn it over for soldering. Bend the leads away from any surrounding pads or lands to avoid shorts during soldering. I try to clip the leads before soldering to avoid mechanical shock to the solder joints.

Once the resistors and diodes are installed, install the IC sockets. Take extra care to install the IC sockets in the proper orientation. The little divot on the end of the socket *must* line up with the one on the PCB legend. It indicates the end of the IC where pin 1 is located. To solder the ICs to the PCB, first clamp a length of thin solder into a "helping hands" alligator clips device (or tape it) so that it protrudes several inches over the edge of your workbench. This is necessary because you'll need one hand to hold the PCB and unsoldered IC socket in place and the other to wield your soldering iron. The bench is playing the part of a third person holding the solder for you.

Now insulate the tip of your forefinger with something that will not conduct heat (some tape, a piece of Post-it note, etc.). Then place the correctly oriented IC socket into the PCB and hold it down with your forefinger while picking up and inverting the PCB with your remaining fingers. Move the inverted board so that the solder in the "helping hands" is positioned near one of the protruding corner leads of the IC socket, and then solder that lead. Do the same for the opposite corner, and then remove your finger, lay the board on the bench, and solder the remaining IC socket pins. I find this easy to do, but you may elect to simply tape the sockets in place, invert the board, and solder them. However, this can leave cooked tape adhesive residue on the socket pin receptacles.

After that, install the capacitors and then the remaining active components. Don't linger with the soldering iron when soldering active components (transistors, diodes, ICs, etc). They are only rated for a few seconds of exposure to soldering temperature. Overexposure to soldering temperature has the potential to change a component's characteristics. When you see solder flow, remove the tip and let the joint cool.

Inspect your solder joints, looking for any that need to be reflowed or any tiny solder balls that might be lurking. Clean the excess flux from the PCB as described in Chapter 2 and dry it with canned air.

Figure 4-7 is a view of the PCB showing the values of each component, which is useful for when you're populating the PCB. With this view, you don't need to constantly go back and forth between the schematic and the silk-screened component designators to determine which component value goes where.

I suggest that you wait until after the PCB is wired to the front panel before plugging the integrated circuits into their sockets. You might as well minimize the chance of damaging them during construction.

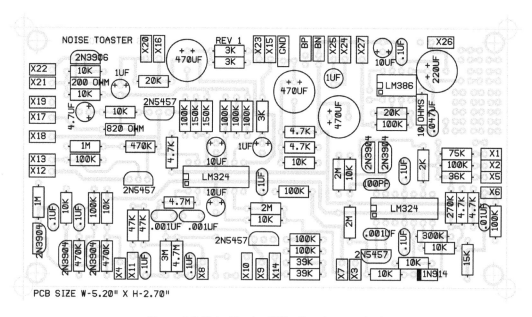

Figure 4-7. *Noise Toaster PCB with values overlaid*

Making the Noise Toaster Front Panel

Figure 4-8 shows a view of the suggested Noise Toaster front panel. When scaled correctly (100%), the front panel template should be a square 6.5" x 6.5". The 100% scaled drawing is also designed to be used as the front panel drill guide. I suggest that you fabricate the front panel out of aluminum, as it provides a good ground plane for all of the potentiometer bodies and the unit's 1/4" output jack. If you don't use aluminum, I suggest you use copper tape or another mechanism to connect the pot bodies to the circuit's ground.

You can download the front panel template (http://bit.ly/YKqX7Y) or buy a professionally machined panel with a silk-screened legend from the Music From Outer Space website.

Print out a copy of the 100% scaled front panel template and use it as a drill guide for your front panel (see Figure 4-8). First, cut the drawing out

along the thin lines around the edge. Next, place the drawing over the aluminum panel on a work bench, aligning it so that it sits parallel to the panel's edges. Tape the drawing in place so that it doesn't slide around while you use a sharp center punch to carefully make drill guide indentations at each of the template hole crosshairs. Next, carefully drill 5/32" holes at all of the indentations, doing your best to stay on the indentations made by the punch. These will serve as guide holes when drilling the holes to the sizes necessary to mount the components. When drilling the holes, I lubricate the bit using a light oil (such as WD-40). It helps the bit cut and reduces burring.

Next drill the holes out to size as follows:

- All mini-toggle switch mounting holes should be drilled to 1/4" (6.4 mm).
- All potentiometer mounting holes should be drilled to 5/16" (8 mm).
- The LED mounting hole should be drilled to 2/10" (5 mm).

- The 1/4" jack mounting hole should be drilled to 3/8" (9.53 mm).
- The push button mounting hole should be drilled to 9/32" (7.14 mm).
- Leave the PCB mounting holes as they are.

After drilling the holes, remove any remaining burrs, clean the panel with soap and water to remove any grease or grime accumulated during drilling, and set it aside.

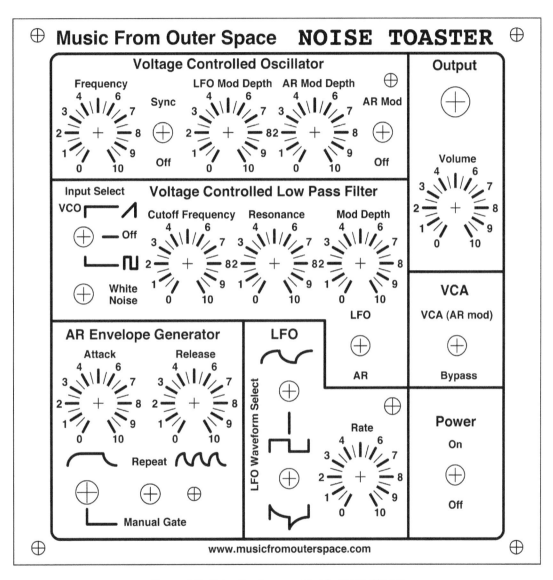

Figure 4-8. *Noise Toaster suggested front panel layout*

Making the Noise Toaster Front Panel Legend Overlay

After the panel is drilled, it's time to make the legend overlay that will be glued to the front panel as a legend for the controls. Once again, print out a 100% scaled copy of the front panel drawing (see Figure 4-8) on whatever color paper you want your front panel to be. Using a heat laminator, apply a plastic laminating jacket to the drawing. This will keep the legend overlay clean and neat looking. Next, carefully cut out the laminated legend overlay, drawing along the thin lines along the edges.

Once your legend overlay is laminated and cut out, place it on the drilled front panel and hold it up to a bright light so you can align the crosshairs of the front panel legend overlay with the centers of the holes on the aluminum panel. Get a good idea of where the panel template will be placed, and then set the panel overlay aside. Next, apply a thin layer of contact cement to the front of the aluminum panel, being careful not to get it in the holes. You don't have to cover the whole surface, but it's a good idea to try and get close to the outside edges of the legend overlay.

Now, before the contact cement dries, place the legend overlay onto the panel as close as possible to where you know it should be. Hold it up to the light again and adjust it by nudging it to where the crosshairs are aligned with the centers of the panel's holes. When it is located properly, carefully burnish the legend template in place by placing a piece of paper over the glued-in-place legend template and gently rubbing it down so that the legend template adheres to the contact cement and the contact cement spreads evenly. Be careful not to move the legend overlay during burnishing.

Place the panel with the attached legend overlay on a flat surface with a large book on top of it, and let the contact cement dry for as long as the package says it should. Once the contact cement is dry, use a hobby knife to carefully cut out all of the holes in the legend overlay (Figure 4-9). Work from the front of the panel with the legend overlay facing you. Poke a hobby knife tip through the holes, and then use the rim of the hole in the aluminum panel to guide the hobby knife as you cut out the holes. Once all of the holes are cut out (including the PCB mounting holes), you are ready to mount the front panel components.

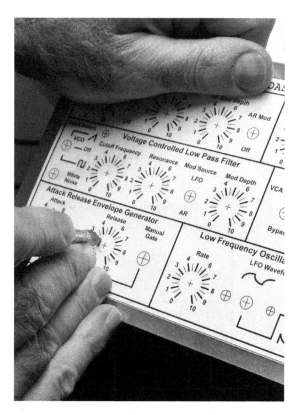

Figure 4-9. *Cutting legend overlay holes*

Installing the Noise Toaster Front Panel Components

The Noise Toaster front panel wiring diagram (Figure 4-10) is a view of the Noise Toaster's controls from the back of the front panel. Start by mounting all of the potentiometers first. Often, potentiometers have a small tab that protrudes from the front that you will need to break off so that the potentiometers sit flush against the panel when mounted. Simply grasp the tab with pliers, bend it, and voila—it snaps right off. Each pot should be mounted to the control panel so that its shaft protrudes from the front of the panel. Secure each pot with the washer and nut that comes with it, observing the orientation shown in the drawing. Tighten all pots to the front panel using pliers or a crescent wrench, being careful not to mar the legend overlay. Attach the knobs to the pot shafts and tighten the set screws using a small jeweler's screwdriver.

Next, mount the switches, observing the orientation shown in the front panel wiring diagram (Figure 4-10). Be careful to mount the right type of switch in each position. The SPDT and the SPDT C.O. type switches look identical, so make sure you don't confuse them during mounting. You can tell the difference between them because the SPDT C.O. switches have a center position, whereas the SPDT switches do not. You can use SPDT switches for SPST switches by simply using the center pole and one of the other terminals. For most mini-toggle SPDT switches (including those sent with MFOS kits), the bat points away from the terminal that is in contact with the center pole. If using SPDT switches instead of SPST switches, mount them as shown in the front panel wiring diagram (Figure 4-10), wire the two terminals shown, and leave the third terminal unconnected.

MFOS NOISE TOASTER Rear Wiring View

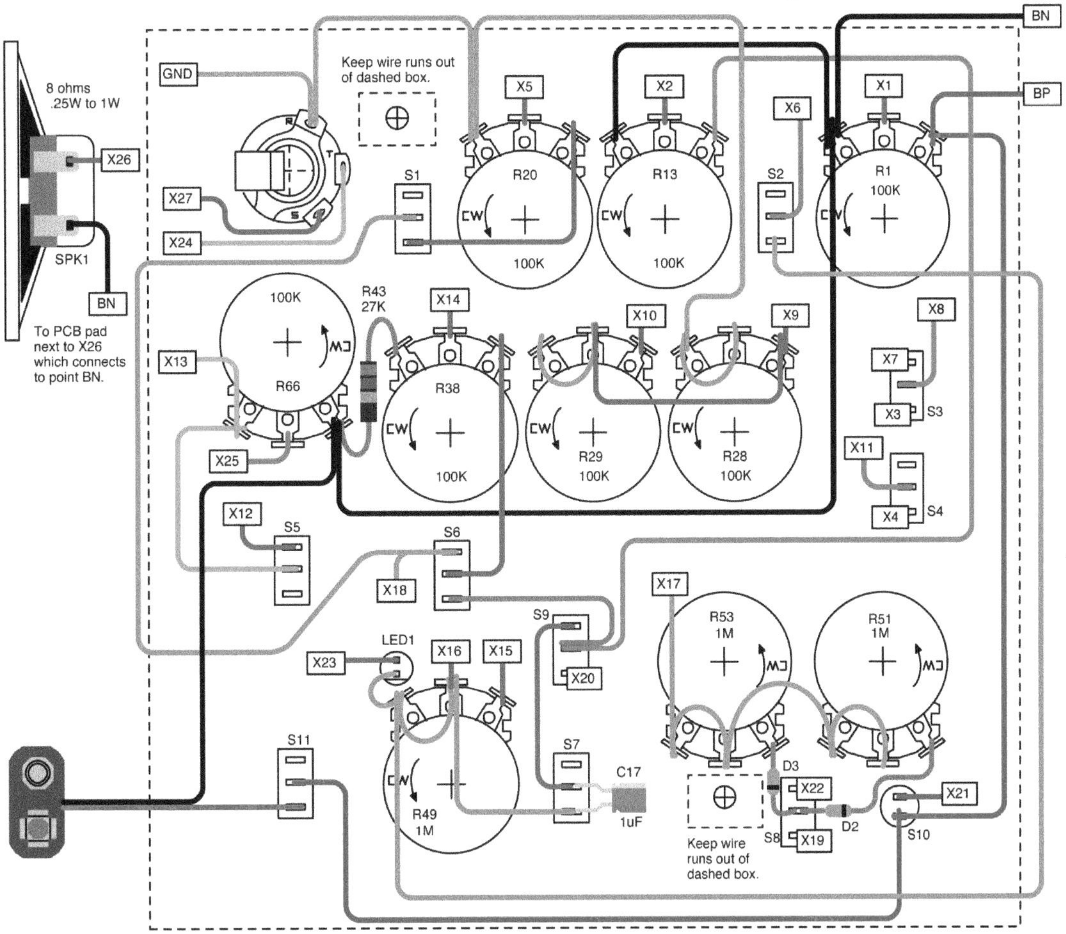

Figure 4-10. *Noise Toaster suggested front panel wiring*

Finally, mount the remaining push button switch (S10), the 1/4" output jack (J1), and LED1. To mount the LED, I cut a small piece of 1/10" (2.54mm) grid perf board with copper-clad do- nut pads into a small rectangle of about 1/4" x 1/2" (6.35mm x 12.7mm). I solder the LED to the middle of the perf board. Then I apply double-stick foam tape to the "wings" and tape the assembly to the panel so that the LED protrudes through the front panel. I find that this method holds the LED to the panel quite nicely. Suppliers also manufacture LED ring clips that hold LEDs to a front panel, which you can purchase if you prefer. You may need to carefully drill the LED hole out a bit to get the ring clip to fit properly. The cathode of LED1 will get soldered to the terminal of R49 as shown. Don't solder it until the other wires connected to the same terminal are installed.

Make sure all of the front panel component mounting hardware is nice and snug and you're ready to move on to wiring the front panel components. Don't use excessive force while tightening the front panel component hardware or you may strip the threads on pots, switches, and/or jacks.

Wiring the Noise Toaster Front Panel Components

Now that the front panel mechanical components are mounted, it's time to do the intrapanel wiring (see Figures 4-10 and 4-11). Wiring mistakes are some of the most frequently encountered problems when building a project. To avoid overlooking a connection, I recommend that you make a copy or two of the wiring diagram to record each wire installed with a highlighter. When you're done, look it over *very* carefully and maybe have a friend look at it, too. Two heads are often better than one, and when you've been staring at something a while, your brain can simply refuse to see something that another person will pick out right away.

The wiring diagram shows additional panel-mounted components (C17, LED1, D3, D2, and R43), the panel-mounted component terminal interconnections, the PCB pad legend labels that

get wired to the front panel component terminals, and the battery's positive and negative connection points. Where the colored wiring indicators end at component terminals, they should be stripped and soldered. However, when a front panel terminal has connections to other front panel component terminals as well as a wire to the PCB, I recommend the following: either leave the terminal unsoldered until you run the wire to the PCB later, or solder it in such a manner that you can still fit the PCB-to-panel wire into the terminal later. It is better to have the PCB-to-panel wires properly attached to front panel component terminals rather than tack soldered to them, as tack soldering provides less physical strength.

Next, you should install the five panel-mounted components (C17, LED1, D3, D2, and R43). You should already have mounted LED1. Connect resistor R43 (27K) between the terminals of pots R38 and R66 as shown. Since the connection on the R38 side has no other connections, you can solder it in place there, but leave the other side unsoldered until the additional wires are added. Capacitor C17 (1µF nonpolar aluminum) is mounted across the terminals of S7 as shown. Diode D3 gets installed so that its anode is connected to the terminal of R53 as shown, and its cathode gets connected to the center pole of S8. Diode D2 gets installed so that its anode is connected to the center pole of S8, and its cathode gets connected to the terminal of R51 as shown. Solder the two diode leads that get connected to S8's center pole, and then solder the other ends to their respective terminals, since no additional wires get connected at either terminal. It is important that the diode leads don't contact anything else.

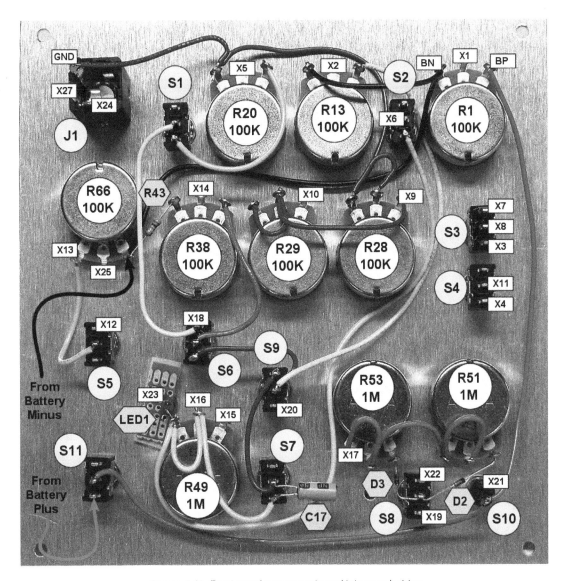

Figure 4-11. *Front panel components and intrapanel wiring*

Make sure the intrapanel wiring is tucked in close to the components so that the PCB that gets mounted on top of them later is not putting pressure on any wires. If a solder joint with a sharp end is pressed against a wire, the wire's insulation may flow away from the pressure point, resulting in a short circuit down the road.

Preparing the Noise Toaster Case

The Noise Toaster's battery holder and 8 ohm 1W speaker (SPK1) both get mounted within the project's case, as shown in Figure 4-12. You can get as creative as you like with your project's case, or you can just use one from a manufacturer like BUD, LMB, or Radio Shack. This case was fabricated using 1" x 3" select pine wood. I buy the 4' length and cut the pieces for the case as needed. I used a dado blade to make the shelf that the front panel and case bottom mount to, but instead you could make the box so that the interior dimensions are 6.5" x 6.5" and then use strips of 3/8" thick pine to create the mounting shelfs. Do this for both the aluminum front panel and for the bottom case covering. I used 1/8" thick masonite for the bottom of the case. I cut the case bottom in half so that I could glue the half where the speaker is mounted, but use screws for the other half to provide battery access.

Figure 4-12. *Project case with speaker and battery holder mounted*

For the speaker grill, I used a piece of 1/16" thick powder-coated perforated aluminum I bought at a hardware store. First I drilled a hole in the bottom of the case about the size of the speaker's cone, and then I glued the perforated aluminum on top of it. After that I glued the speaker on top of the grill material. I used contact cement for mounting both.

I mounted the battery holder on the inside of the case, and as mentioned above, half of the bottom of the case on the battery's side can be removed to install or change the battery. The battery holder in the component list is from Radio Shack, but as you can see, any 9V battery holder will do.

Be careful to align the speaker so that it is centered on the hole you drilled for it. After contact-

cementing the grill and speaker into the case, leave it undisturbed to dry for the time recommended on the contact cement package. When the contact cement has dried, twist a pair of 8" length 22 or 24 gauge stranded wires together and carefully solder them to the speaker's terminals.

I used a set of clear polyurethane bumpers for the *feet* of the unit (Figure 4-13). The ones I used are 0.8" (20.32 mm) in diameter and 0.4" (10.16 mm) tall. They come on a sheet of 24 and peel off individually to expose their adhesive backing. They stick really well and hold the unit up nicely so that the bottom-mounted speaker is not blocked. I got mine from *All Electronics*, catalog number RF-32.

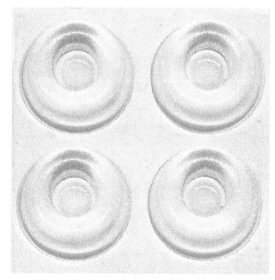

Figure 4-13. *Case bottom rubber feet*

Wiring the PCB to the Front Panel

At this point, it is time to wire the front panel controls to the PCB's external connection pads. All of the external connection pads on the PCB have the format Xn (except BN, BP, and GND). The wiring diagram (Figure 4-10) shows where each wire must be run to the PCB for each Xn point.

When installed, the PCB will be mounted to the aluminum panel, with the electronic components facing away from the front panel using two 1 1/4" long 6-32 machine screws. There are two holes on the front panel that are used for mount-

ing the PCB. Figure 4-14 shows how to orient the PCB on the front panel when mounted.

Figure 4-14. *PCB mounted to front panel*

As long as you get a wire from the panel to the PCB for each Xn position, you'll be good to go, but here is some advice that will help you get it put together correctly. I ran all of the wires to the board from one side to facilitate easy removal of the board for troubleshooting or modifications. The wires have what is called a *service loop*, which is just a fancy name for a bit of slack to

allow the board to be folded back from the panel after wiring, if necessary.

The next step is a bit tedious but it will ensure that you get everything wired up correctly. Do not mount the PCB yet, but visualize where it will eventually be mounted. Attach a length of wire to each Xn point on the front panel that will be long enough to solder to the PCB after it is mounted. Leave adequate length, as you can trim them later before soldering them to the PCB's Xn pads.

As you attach and solder each wire to its front panel control terminal, label the opposite end of it with its Xn designator. The multicolored adhesive-backed dots sold at office supply stores work well for this. Continue soldering and labeling the remainder of the panel-to-PCB wires.

When you believe you have attached and soldered every last one, inspect every terminal on the front panel that the wiring diagram indicates should have an Xn wire attached. Install any missing wires, and reflow any solder joints that appear dull and gray. Now it's time to mount the populated PCB and attach the interconnecting wires.

The PCB gets mounted on the two 1 1/4", 6-32 screws that you previously installed, but you must put two 3/8" long spacers on each of them to "stand off" the PCB first (Figure 4-15). Once the PCB is mounted on the screws and spacers, attach the remaining 6-32 nuts to hold it in place.

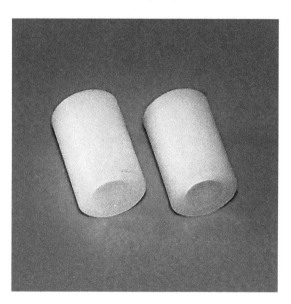

Figure 4-15. *Nylon spacers for PCB stand-offs*

Carefully connect the labeled Xn wires attached to the panel components to the PCB *one at a*

time. Take extra care to connect each Xn panel wire to its corresponding Xn PCB connection point. You can solder the wires to the PCB from the top easily thanks to the top annular ring (donut pad) each plated through-hole has. One at a time, remove each wire's label, and strip about 1/8" to 3/16" of insulation from its end. Place the stripped end into the PCB's Xn pad, leaving enough bare wire exposed on the top side of the PCB to touch with the tip of your soldering iron. Apply the solder iron to the exposed wire and top annular ring simultaneously and feed in some solder. The solder will flow into the stranded wire and the hole and will make fillets on the top and bottom sides of the PCB. Continue to work your way around the board, soldering all of the Xn panel-to-PCB interconnection wires until complete.

Connect the red and black wires of the battery snap to the points indicated on the front panel diagram. The red lead goes to one side of the power switch (S11), and the black lead goes to R66's terminal, as shown in the wiring diagram.

The two wires you attached to the speaker go to the PCB at point X26 and the donut pad right next to it. The pad next to X26 is routed to BN. The speaker's polarization is not important in this application, so simply attach the two speaker wires to the two pads. Figure 4-16 shows the correct pads to attach the speaker wires to outlined in red.

Don't insert the ICs into their sockets yet!

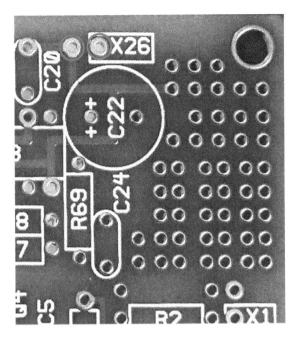

Figure 4-16. *PCB speaker connection points*

Congratulations are in order, because at this point you have completed the construction of the Noise Toaster, except for attaching the front panel to the case and inserting the ICs into their sockets. *Before* attaching the front panel to the case, however, you need to test the unit and, if necessary, troubleshoot.

Testing the Noise Toaster

I'm going to approach the Noise Toaster test procedure under the assumption that each test passes with flying colors. However, if you run into issues, I suggest that you reread the section "Troubleshooting Tips" on page 31 from Chapter 2. The information will give you a lot of possibilities to consider while you're searching for whatever problem stands between you and making cool analog synthesizer sounds. Additionally, at the end of this test procedure I've included the section "My Noise Toaster Is Being Sullen and Listless" on page 82. It contains a table that lists a variety of possible problems and the active components that may be involved with each. You will still need to troubleshoot, but at least you'll have

some signposts pointing you in the right direction.

Before installing the ICs or attaching the battery, use your multimeter to make sure that the battery connector is correctly connected to the unit. Set the meter for the lowest ohmic range, touch the leads together, and remember the reading. Turn the power switch on and put one of the meter's leads on the battery snap's positive lead. Figure 4-17 shows which of the battery snap connections is positive (the other is negative).

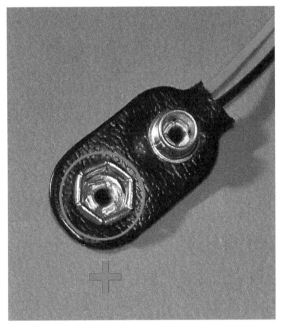

Figure 4-17. *9V battery snap polarity*

Place the meter's other lead on pin 4 of IC1's socket. The meter should read the same as when you touched the leads together. If the reading varies by an ohm or so, no worries, but if it varies by 5 or more ohms, you should revisit the IC socket solder joints and the solder joints for the battery and power switch.

Do the same for IC2's pin 4 and then for IC3's pin 6. With one lead on the battery snap's positive connector, each of the readings should be the same as when you touched the leads together.

Now place one meter lead on the battery's negative connector and the other on pin 11 of IC1. Again, you should see the same low reading as when the leads were touching. Do the same for IC2's pin 11 and then for IC3's pin 4. If you see a significant difference in the readings, you should revisit the IC socket soldering and the battery wire soldering.

If all is well, turn off the power switch, insert the ICs into their sockets, *making sure to observe the correct orientation*, and connect a fresh 9V battery to the battery snap. Support the board with your fingers while applying the pressure required to insert the ICs. Set the front plate on the project case during testing, but don't attach it with the screws that came with it yet.

Testing the White Noise Generator

1. Set the Noise Toaster's controls as shown in Figure 4-18. The switch icons show which way to set the toggle switches. For two position switches (SPST and SPDT), there are two states down (bottom box filled in) and up (top box filled in). For SPDT C.O. (center off type switches), there are three states down (bottom box filled in), up (top box filled in), and centered (middle box filled in). Set knobs to the position with the red indicator.

2. Now turn the power switch to on. After a second or two, you should be hearing white noise coming from the speaker. The short delay after powering the unit is the time needed for the virtual ground capacitor C15 to charge up. If you turn the Output Volume control up, the white noise should get louder, and if down, softer. Turn the White Noise switch to the down (or off) position, and you should no longer hear it. Leave the controls in this configuration.

If that test went well, it means that the white noise generator works, the 1W amplifier and speaker works, the volume control works, and the White Noise switch is wired correctly. Bravo!

Testing the VCO

To prepare for this test, return the unit's controls to the settings shown in Figure 4-18 again. Set the VCF's White Noise Input Select switch to Off.

1. Set the VCF's Input Select Switch for the VCO to the Ramp Wave position (up). You should be hearing the ramp wave coming through the speaker. Set the VCF's Input Select Switch for the VCO to the Square Wave position (down). You should be hearing the square wave coming through the speaker with a more hollow timbre.

2. Return the VCF's Input Select Switch for the VCO to the Ramp Wave position (up). When the VCO's Frequency knob is turned clockwise, the VCO's frequency should increase. Depending on your hearing's frequency range, you may still hear it when it is turned all the way up. When you turn the VCO's Frequency knob the other way, it should go down in frequency until you can only hear clicks through the speaker. If you turn the knob too far counterclockwise, the VCO may stop oscillating (which is fine).

3. Return the VCO's Frequency knob to the twelve o'clock position. Turn the VCO's LFO Mod Depth control up slowly. The VCO's frequency should begin to go up and down in time with the LFO's flashing LED rate indicator. As the VCO's LFO Mod Depth control is advanced, you should hear more modulation and an increase in the VCO's frequency.

4. Return the VCO's LFO Mod Depth pot to the fully counterclockwise position. Set the VCO's AR Mod switch to the AR Mod position and its AR Mod Depth pot to the twelve o'clock position. The frequency of the oscillator will fall and may even stop oscillating (again, it's fine). Now press and hold the Manual Gate button. The VCO's frequency should rise to a maximum and stay there until you relase the button, at which time the VCO's frequency should fall back to where it was.

5. Return the VCO's AR Mod Depth pot to the fully counterclockwise position and its AR Mod switch to the Off position. Turn the LFO Rate pot fully clockwise. Set the VCO's Sync switch to the Sync position. The tone of the VCO should change. Now advance the VCO's Frequency pot slowly upward. The VCO's timbre should change in a sweeping manner, emphasizing harmonics as it does so.

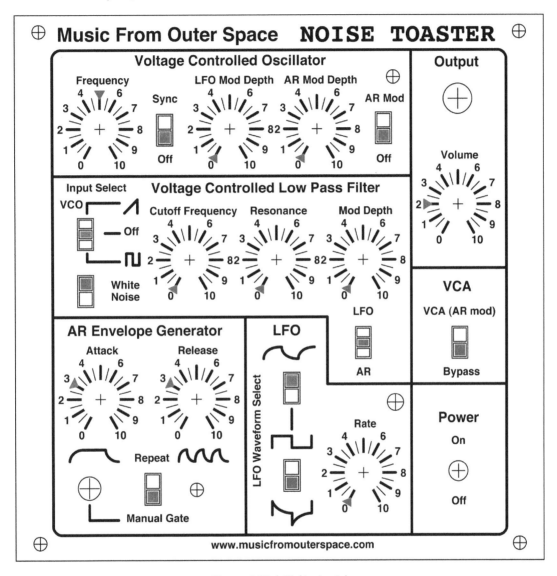

Figure 4-18. *Initial test patch*

If everything worked, you deserve a pat on the back before continuing. The VCO works properly, the LFO is oscillating, the AREG is functional in Manual mode, and the Manual Gate push button works.

Testing the VCF

To prepare for this test, return the unit's controls to the settings shown in Figure 4-18 again. The VCF's White Noise Input Select switch should be set to On.

1. You should be hearing white noise coming through the speaker. Slowly turn the VCF's Cutoff Frequency knob down. The cutoff frequency of the filter should go lower as you do so. Leave the Cutoff Frequency knob in the counterclockwise position.

2. Set the Resonance knob to its fully clockwise position. Slowly turn the VCF's Cutoff Frequency knob up. The cutoff frequency of the filter should go higher as you do so, and the sound should be more howling in nature. Play with the Cutoff Frequency knob and make some wind sounds for a while. Adjust the resonance down a bit and compare the sound. If you've got good cutoff and resonance adjustment, excellent.

3. Set the White Noise Input select switch to Off. Return the Resonance knob to its lowest setting and the Cutoff Frequency knob to its highest setting. Set the VCF Input Select VCO Switch to Square Wave, and adjust the VCO's Frequency knob until its frequency is very low. Slowly turn the VCF's Cutoff Frequency knob down. The cutoff frequency of the filter should go lower as you do so. Leave the Cutoff Frequency knob in the counterclockwise position.

4. Set the Resonance control to its fully clockwise position. Slowly turn the VCF's Cutoff Frequency knob up. The cutoff frequency of the filter should go higher as you do, accompanied with an opening wah-type sound. Adjust the VCO's Frequency control down until it is only ticking. Play with the Cutoff Frequency control and make some cool ticking or popping sounds for a while.

5. Turn the Resonance and Cutoff Frequency controls completely down again. Set the VCF's Mod Source switch to the AR setting and the Mod Depth control to twelve o'clock. Press the AREG's Manual Gate button. The filter's cutoff frequency should rise while you hold the button down and then fall when you release it.

6. Set the AREG's Repeat/Manual Gate switch to Repeat. The VCF's cutoff frequency should be modulated by the AREG's repeating envelope. Turning the Mod Depth control up should increase the modulation, and turning it down should decrease it. The Mod Depth knob must be turned beyond 2 or 3 for the effect to be heard.

7. Adjust the LFO's Rate control to 7. Switch the VCF's Mod Source switch to LFO and its Mod Depth control fully counterclockwise. Advance the Mod Depth control to beyond 6, and you should hear the VCF's cutoff frequency being modulated by the LFO's Integrated Square Wave. As you advance it further, the modulation should intensify. Switch the LFO's Waveform Select controls to select the Square wave output. Now adjust the VCF's Mod Depth control to below 5, and you should hear its cutoff frequency being modulated by the LFO's Square Wave output.

If everything worked, you deserve a beverage break before continuing. The VCF works properly, and the AREG's Repeat functionality works, too.

Testing the VCA

To prepare for this test, return the unit's controls to the settings shown in Figure 4-18. Set the VCF's White Noise Input Select switch to Off.

1. Set the VCF Input Select VCO Switch to Square Wave, and adjust the VCO's Frequency knob to twelve o'clock. Turn the VCA's select switch to VCA (AR mod). Press the Manual Gate button on the AREG. You should hear the VCO slowly come on and stay on as long as you hold the button. When you release the Manual Gate button, the VCO should fade away.

2. Set the AREG's Repeat/Manual Gate switch to Repeat, the Attack control to 0, and the Release control to 1. You should be hearing

the VCO with a repeating volume envelope kind of like *"bink-bink-bink-bink…"*

If everything worked fine, do a couple of victory yoga poses as a reward before continuing. The VCA is working great.

Testing the AREG

If the other tests you've been doing have all worked as described, then the AREG is functioning properly.

Testing the LFO

To prepare for this test, return the unit's controls to the settings shown in Figure 4-18 again. Set the VCF's White Noise Input Select switch to Off. Set the LFO's Rate to 8.

1. Set the VCF's Input Select Switch for the VCO to the Ramp Wave position (up). You should be hearing the ramp wave coming through the speaker. Advance the VCO's LFO Mod Depth control to 5. You should hear the VCO's frequency sliding up and down. Advance the LFO's Rate knob; the frequency of the modulation should increase, and the LFO's rate indicator LED should flash in sync with the modulations. When the LFO's Integrated Square Wave is selected, the depth of the modulation will decrease as the LFO's Rate control is advanced.

2. Reduce the VCO's LFO Mod Depth to 2. Change the LFO output waveform to the Square wave (LFO's top switch down, bottom switch up). Since the LFO's Square wave output has greater amplitude than the Inte-

grated Square Wave, it produces more modulation.

3. Increase the VCO's LFO Mod Depth to 5. Change the LFO output waveform to the Differentiated Square Wave (LFO's left switch down, right switch down). The character of the modulation should change to sound more like a bird chirping.

Testing the Output Jack

When you insert a phone plug into the output jack, the internal amplifier and speaker are bypassed, and you should not hear the unit's speaker. The volume control should continue to control the unit's output level when it is connected to an amplifier or mixing board.

If everything worked, find someone to nominate you for *person of the year* before continuing. Your Noise Toaster is ready for prime time! Insert the battery into the battery holder and attach the panel to the project case with its included screws. Watch that you don't pinch the speaker or battery wires while attaching the panel.

If your Noise Toaster checks out and you want to get straight to making cool sounds, you can move on to Chapter 7. I heartily encourage you to come back and read Chapters 5 and 6 when you want to learn exactly how all those cool sounds are being made.

Figure 4-19 is a blank Noise Toaster patch sheet with which you can document the control settings when you find an elusive sound you want to reproduce at a later time.

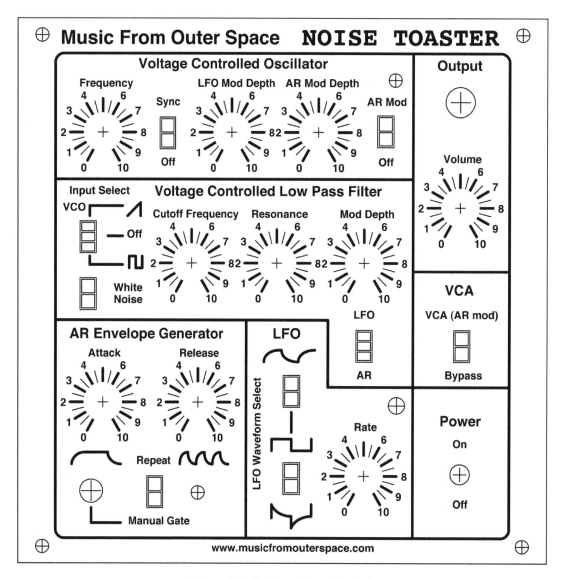

Figure 4-19. *Noise Toaster patch sheet*

My Noise Toaster Is Being Sullen and Listless

It's a bummer when something doesn't work right the first time, but if it's any comfort, we've all been there. I first want to assure you that if the project is built according to the plans presented, it will absolutely work. When all of the correct component values are where they should be and *all of the active components are good*, your Noise Toaster will work like a charm. If one or more of

the tests above failed, it's time to put on your thinking cap and start troubleshooting.

As I mentioned at the outset of the test section, it would be a good idea to read the section "Troubleshooting Tips" on page 31 from Chapter 2 to get some good ideas of what can go wrong with a project before beginning to troubleshoot. You're going to have to trace the operation of the circuit, following the schematic and probing around until you find something that is acting strangely.

Check your panel wiring for missing or misrouted wires. Check the component values on the PCB, looking for incorrect values or missing components. Look back over your soldering, and if a solder joint looks suspect, give it some reflow love. Inspect the project with a magnifying glass, looking for tiny solder balls that may be shorting adjacent pads, lands, or component leads.

I have found both the LM324 and LM386 to be hardy and forgiving chips, but applying reversed power or delivering a twitch-inducing static shock to either of them can destroy them, so you may need new chips if nothing seems to work and the wiring is all there and the components are all correctly oriented and soldered.

Table 4-3 lists a variety of problems you may encounter with the unit's modules and the active components that may be involved if they are not working properly. Besides a bad active component, a dead battery, incorrect wiring, poor soldering, missing components, or incorrect component values are pretty much the only remaining possibilities.

Table 4-3. Possible issues and associated active components

Problem	Associated Active Components
No white noise output	Q2, Q5, Q6, U2, U3
No VCO ramp output	U1, Q4, Q3, Q1, D1
No VCO square output	U1, Q4, Q3, Q1, D1, U2
VCF not working	U2, Q8
VCA not working	Q7
LFO not working	U2
AREG problem	U2, Q9, Q10, D2, D3
1W amp problem	U3

The Incredible Op Amp 5

You may be familiar with operational amplifiers (or op amps) and have perhaps even done some experiments with them. They are one of the most useful circuit elements you'll encounter, and it's well worth the effort to become proficient in applying them in your synth-DIY circuit design and experimentation work.

The Noise Toaster design uses several op amps for different purposes; before we dissect the various circuits of the Noise Toaster, it is important to have a good idea how op amps work. I'm not going to cover the chemical formula for cadmium yellow here. Instead, I'm going to show you how to paint things; specifically I'm going to show you how to paint circuits with op amps.

Figure 5-1. *Typical op amp IC packages.*

There are hundreds, perhaps even thousands of unique op amp ICs. Figure 5-1 shows some typical op-amp packages. Op amps can be powered from as little as +/–1.5V or as much as +/–18V, and there are special op amps deigned to operate from a single voltage supply. Op amp ICs contain one, two, or four independent op amps and generally come in 8- or 14-pin DIP (dual in-line package) or SOIC (small outline integrated circuit) packages. Every op amp has a corresponding data sheet detailing all of the op amp's

parameters (which are legion) and also, importantly, which pin is which. Op amps have inverting inputs, noninverting inputs, outputs, power connections, bias connections, offset adjustment connections, etc., which if not connected properly can result in baked op amp (not a favorite dessert of anyone I know).

Op Amps as Amplifiers

Op amps can perform many functions, but one of the most common is amplification of low level signals. A typical op amp can easily provide gains of up to 1,000, and the op amp parameter known as *open-loop gain* can be as high as 1,000,000. However, you need to understand that when I say "amplify," I'm not talking about the kind of amplification you get from your guitar amplifier. You can't connect your guitar up to an op amp with a gain of 1,000,000 and hook the op amp's output to a giant speaker cabinet and rock the universe!

You *can*, however, connect your guitar to an op amp circuit whose gain is between, say, 10 and 100, and boost its output to line level (about a volt or so AC). Op amps are not power amplifiers. Power amplifiers that drive speakers generally have large heat-sinked power transistors or integrated IC amplifiers (also containing power transistors) to do the heavy lifting. But you can be sure that op amps *will* be found in the front end of your amplifier, providing the initial gain for the low level signals connected to its input, be they microphones or guitars. And it's a safe bet that op amps will be in the active filtering circuits used in the amplifier's front end tone controls as well. Let's go further into what makes op amps so useful.

Negative Feedback Demystified

Op amps use what is known as *negative feedback* to produce precise amounts of gain (amplification), and they make it as easy as pie to get whatever gain you need. Figure 5-2 shows the circuit that you could use to amplify a low-level signal (a guitar or microphone for example) up to a level that would be considered *line* level. In this section I'll try to demystify op amps and negative feedback and then show you some practical uses for them in your own synth-DIY work.

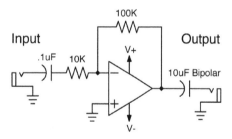

Figure 5-2. *Op amp inverting buffer*

operation. Some op amps have input impedance as high as 10^{12} ohms, which means they consume very, very... very little current. Thus, when you buffer a signal with a high-impedance input op amp, the circuit being buffered hardly notices that the op amp input is there at all since it consumes so little current. This is one of the reasons you need to use a high-impedance input op amp in sample and hold circuits. The cap holding the sample voltage does not get discharged by the high-impedance op amp buffering it. The output of the op amp is just the opposite. The lower its output impedance, the more current drive capability it has.

In Figure 5-2, we see the op amp has an input resistor value of 10K and a feedback resistor value of 100K, which results in an inverting amplifier (or buffer) with a gain (or amplification factor) of 10. The term *buffer* refers to the op amp's ability to take a low-level, often high-source impedance signal and amplify it, as well as provide the amplified signal from its relatively low-impedance output with some current drive capability. We're talking milliamps here—nothing grand—but within most electronic circuits that's all we need to do whatever we want. Before we can understand what's happening in this circuit, we need to consider Figure 5-3.

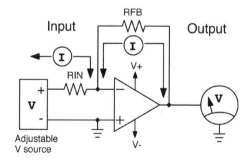

Figure 5-3. *Understanding negative feedback*

We are going to discuss this circuit's response to DC to understand how negative feedback determines the gain of the circuit. For our demonstration, the input of the op amp is connected to the output of the adjustable voltage source, and the output of the op amp is connected to a voltmeter. One goal of op amp designers is for the op amp inputs to be high impedance so that they have minimal loading effect on signals applied to their inputs. The impedance of op amp inputs can be as high as one gigaohm (1,000,000,000 ohms)! When configured as an inverting buffer, the op amp's internal circuitry attempts to keep its inverting input (marked with a minus symbol) and its noninverting input (marked with a plus symbol) at the same voltage by adjusting its output voltage.

The adjustable voltage source is applied to one side of the input resistor RIN, whose other side is connected to the inverting input (-) of the op amp. The feedback resistor (RFB) is connected between the op amp's inverting input and the op amp's output. The values of both RIN and RFB determine the gain (or amplification factor) of the op amp circuit. Consider that RIN is 10K and RFB is 20K, and remember the op amp's job (when configured as an inverting buffer) is to keep the inverting and noninverting inputs at the same voltage by means of the feedback resistor. If we set our adjustable voltage source to 1V, what will the op amp's output do? What voltage is the noninverting input at? It is at 0V (GND), so the op amp's internal circuitry is going to try and keep both inputs at 0V.

*It's not hard to put this circuit on a breadboard and experiment with it yourself. **Figure 5-4** shows a solderless breadboard with this experiment wired up. Power is supplied by two 9V batteries. One battery is for the positive supply, and the other for the negative, and the remaining positive and negative leads are connected to create the circuit's ground. You can open the input and/or feedback circuit and insert an ammeter to observe the currents flowing in the circuit. The trimpot supplies the adjustable voltage for the experiment.*

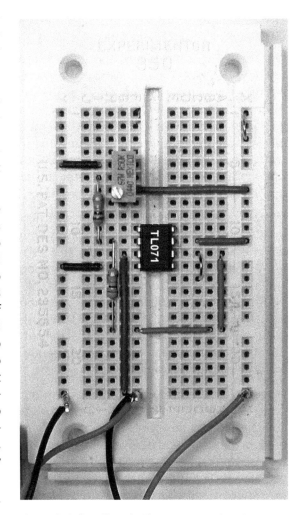

Figure 5-4. *Breadboard with op amp experiment*

The op amp keeps both inputs at the same voltage by adjusting how much current is pushed or pulled through the feedback resistor from the op amp's output. In this case, with 1V fed to the inverting input via our 10K RIN resistor, the op amp's output is going to have to go to −2V in order to pull enough current (100µA) through the feedback resistor to balance the current arriving through the input resistor. Since we have 100µA arriving at the inverting input and 100µA flowing away from it, the net current is 0, and the voltage at the inverting input is 0V. Notice that with the 10K input resistor and the 20K feedback resistor, 1V of input voltage got translated into –2V on the output of the op amp. We say that the op amp exhibited a gain of minus 2.

What if we adjust the voltage applied to RIN to 2V? Now we'll have 200µA flowing to the inverting input through the 10K input resistor. To keep the inverting input at 0V, the op amp's output is going to have to change its voltage so that it pulls 200µA away from the inverting input through the feedback resistor RFB. To pull away 200µA through 20K, we need –4V, and the output of our op amp obliges and adjusts itself accordingly.

How about if we adjust the voltage applied to RIN to –2V? Now we'll have 200µA flowing away from the inverting input through the 10K input resistor. To keep the inverting input at 0V, the op amp's output is going to have to change its voltage so that it pushes 200µA toward the inverting input through the feedback resistor RFB. To push 200µA through 20K, we need 4V so that the output of our op amp behaves like a perfect gentleman and delivers the goods.

Op amps are amazing, and without them our synth-DIY work would be about 20 times more complicated, circuitry-wise. They contain hundreds of transistors and passive components like resistors and capacitors (albeit with small values of C) as well. Op amps of today are precise despite their low cost, and when it comes to determining gain, we don't have to hook up various values of resistance and then check to see if we got it right. There is a simple formula used to calculate the gain for an op amp used as an inverting buffer, shown in Figure 5-5.

Inverting Buffer Gain Formula

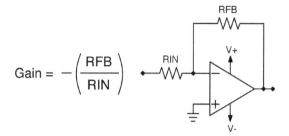

$$\text{Gain} = -\left(\frac{RFB}{RIN}\right)$$

Figure 5-5. *Inverting buffer gain formula*

When using the op amp as an inverting buffer, the gain is simply the value of the feedback resistor divided by the value of the input resistor times –1. Remember that when we applied a positive voltage to the input of the inverting buffer, the output voltage had to go to a negative voltage to balance the effect of the current flowing into the inverting input via the input resistor. Conversely, when we applied a negative voltage to the input of our inverting buffer, the output voltage had to go to a positive voltage to balance the effect of the current flowing out of the inverting input via the input resistor. That's where the *times –1* comes from in the simple gain formula.

Unwanted Op Amp Oscillation

If you discover that your inverting buffer with gain higher than 1 is adding a high-frequency oscillation to its output, it could be because the op amp is not internally compensating for the gain you're using. This oscillation can vary in frequency depending on the op amp, but it is typically above 100 kHz. You will need an oscilloscope to see it, but often you can hear it as excessive noise in the op amp's output. Most op amps are internally compensated but only for a gain of 1. A trick to keep your inverting buffers quiet is to add a 10pF to 100pF capacitor in parallel with the feedback resistor. This lowers the gain of the pesky high frequency enough to get it to shut up.

The Noise Toaster uses an inverting buffer in its VCO as well as in its VCF. In the VCO's case, an inverting buffer (U1-C) is used to mix the control voltages applied to the VCO to modulate its frequency. The op amp continually adjusts its output voltage in response to the current arriving at its inverting input from the three input sources. The output voltage then controls the voltage to the current converter comprised of U1-D and its associated components.

The Noise Toaster's VCF uses an inverting buffer (U2-A) to both mix the signal sources applied to its input and to apply filtering to the signal via the additional components found in its negative feedback network.

This is all well and good, but what if I don't want the output to be inverted? What if I want gain but that's all I want? The friendly op amp is ready to serve. Lets consider the op amp as a noninverting buffer, as shown in Figure 5-6.

Noninverting Buffer Gain Formula

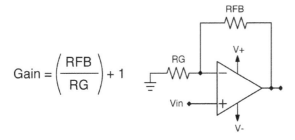

$$\text{Gain} = \left(\frac{RFB}{RG}\right) + 1$$

Figure 5-6. *Op amp noninverting buffer*

The op amp works in the same way that it did before, except we've configured things a bit differently. We still have a feedback resistor (RFB) as before, but instead of applying the input to the inverting input via a resistor, we simply connect the inverting input to ground through a resistor. We have named this resistor to ground RG. Another difference to note is that we are applying the signal we want to amplify to the noninverting input. Let's consider applying a DC voltage to the noninverting input and see what voltage the output of the op amp needs to attain in order to keep the inverting input and the noninverting input at the same voltage.

For our example, we will apply values of 10K for RG and 39K for RFB and apply 1V to the noninverting input. Typically, the impedance of an op amp input is high enough that it doesn't source or sink any appreciable current, which is why we ignore its effect on the voltage at the junction of RG and RFB. We know that the op amp will adjust its output voltage to cause the inverting and noninverting inputs to be at the same voltage, so we automatically know that the output is going to drive enough current through the series combinaton of the feedback resistor and the resistor to ground to drop 1V across 10K resistor RG.

Using Ohm's law, we determine that the current required to drop 1V on a 10K resistor is 1V/10K, or 100µA. We know that the current in a series circuit is the same throughout, so there is also 100µA flowing through the feedback resistor. Ohm's law helps us determine the voltage required to drive 100µA through 49K (10K + 39K). It tells us that E (voltage) equals I (current) times R (resistance). Thus, to drive 100µA through 49K, we need (.0001 x 49000)V or 4.9V. So the output voltage adjusts to 4.9V, which is exactly what we expected, since the formula for the noninverting buffer's gain is simply the feedback resistor value, divided by the resistor to ground value, plus 1. (Written as an equation, (39K/10K) + 1, which comes out to 4.9.) An important point to remember when using the op amp in a noninverting configuration is that the op amp's output voltage follows the sign of the input voltage instead of inverting it, as happens in the inverting buffer.

We've discussed DC voltages and current flow to simplify our discussion of op amp negative feedback, but the op amp configured as an inverting or noninverting buffer works the same way when dynamically changing voltages are applied to the appropriate input. The output constantly adjusts its voltage to keep the inverting and noninverting inputs at the same voltage by pumping the appropriate current from the output through the feedback resistor. So returning to Figure 5-2, we now understand that with the input resistor of 10K and the feedback resistor of 100K, the op amp provides a gain of –10. Since we block any DC that might be present in the input signal with the .1µF cap, the op amp only responds to the dynamically changing AC signals arriving at its noninverting input via the 10K resistor RIN.

Op Amps as Comparators

What would happen if we took away the feedback resistor in the op amp circuit? How will the op amp keep the inverting and the noninverting inputs at the same voltage? What kind of crazy world are we living in? Settle down: it's fine if we take away the feedback resistor, but we normally would not use the op amp for amplification without one. When we take away the feedback resistor, the op amp changes its hat to become a *comparator*. As the name implies with the root "compare," comparators monitor the voltage levels of both inputs (inverting and noninverting) and do something when one is higher or lower than the other.

When an op amp is configured as a comparator, it "detects" that one input is higher or lower than the other and adjusts the output voltage to as high or low a level as it can in a vain effort to get the op amp's inputs to be at the same voltage. Without a feedback resistor, there is no way for the output to change the voltage on the op amp's inputs, so it flails wildly to its positive or negative maximum output voltage. The op amp's maximum positive or negative output voltage is generally a volt and a half away from the op amp's positive or negative power supply voltage, although some specialty op amps can come closer to the supply rails in operation. When an op amp's output is as high or as low as it can go, we say it is in positive or negative saturation, respectively.

I'd like to draw your attention to Figure 5-7.

Op Amp Noninverting Comparator

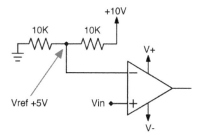

Figure 5-7. *Op amp noninverting comparator*

Here we are applying a 5V reference voltage (Vref) to the inverting input of the op amp, and the noninverting input is the input to the comparator. If we apply a voltage lower than 5V to the noninverting input, the output of the op amp will go as low as it can. The op amp *really* wants to pump current away from the inverting input through the nonexistent feedback resistor. As long as the voltage on the noninverting input is lower than the voltage on the inverting input, the op amp's output will remain saturated low (stuck at its maximum negative output voltage). If the voltage on the noninverting input goes above 5V, the output of the op amp will become saturated high (stuck at its maximum positive output voltage) again in a vain effort to pump current to the inverting input through the nonexistent feedback resistor.

Noninverting comparators are important circuit elements in the Noise Toaster. The VCO uses a noninverting comparator (U1-B) for the ramp core reset circuit. When the output of U1-A ramps beyond U1-B's threshold, U1-B's output shoots high, turning on Q1 and resetting the integrator made up of U1-A and associated components. Op amp U2-B also functions as a noninverting comparator and squares up the ramp signal applied to its input, resulting in the unit's square wave source.

We can also apply Vref to the noninverting input and the comparison voltage to the inverting input to arrive at an inverting comparator, as shown in Figure 5-8.

Op Amp Inverting Comparator

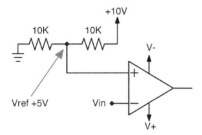

Figure 5-8. *Op amp inverting comparator*

This circuit works much like the op amp noninverting comparator, except that we've switched which input is held steady at Vref and which input is the variable input voltage. As you would expect, when the voltage on the inverting input is higher than the voltage on the noninverting (Vref) input, the output of the op amp goes to negative saturation. If the voltage on the inverting input is lower than the voltage on the noninverting input (Vref), the output of the op amp goes to positive saturation.

An inverting comparator is used in the Noise Toaster's attack release envelope generator's (AREG) auto-repeat circuit. When the voltage on C18 (4.7µF electrolytic capacitor) is below the threshold voltage applied to the noninverting input of U2-D, the output of U2-D is saturated high, providing current to charge C18 via 1M Attack Time pot R51 and 1N914 diode D2. When C18 charges to a voltage level that exceeds the threshold set by R64 and R63 (both 150K resistors), U2-D's output shoots low, providing a path for C18 to discharge via Release Time pot R53 and 1N914 diode D3. When C18 discharges enough for the voltage on U2-D pin 13 to go below the voltage on its noninverting input, its output shoots high again, and the cycle continues, resulting in repeated AR cycles. The 100K resistor R58 provides a bit of positive feedback (and resulting hysteresis), which is critical to the circuit's operation. Without the hysteresis zone provided by R58, the amplitude of the auto-repeating AR

output would be extremely low and the repeating frequency much higher.

Positive Feedback (Hysteresis)

Op amp comparators can be really finicky when the voltages on the inputs are very close to each other. Op amps operate in their open-loop gain mode when used as comparators (remember that open-loop gain can be as high as 1,000,000). Thus, when the voltages on the op amp's inputs are within a few millivolts of one another, any noise in the circuit can cause one or the other input to be spuriously modulated, causing it to rise and fall in voltage above and below the threshold voltage. This can cause the op amp to flail wildly between positive and negative saturation. This has the capacity to cause serious issues in certain applications, so we should see if there is something that can be done to alleviate this problem. *Positive feedback* comes to the rescue. Let's consider the circuit in Figure 5-9.

Op Amp Noninverting Comparator with Hysteresis

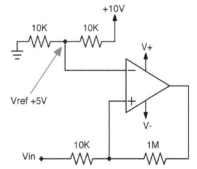

Figure 5-9. *Op amp noninverting comparator with hysteresis*

Here we see that we have added a *positive feedback* resistor between the op amp's output and its *noninverting* input. We have also added a resistor between the applied input voltage and the noninverting input. Both of these resistors are necessary to achieve hysteresis and alleviate the output chatter that occurs when the op amp's differential inputs are close together in voltage. With the positive feedback resistor in place, the output of the op amp is reinforcing the level on the noninverting input, whether saturated high or low. For example, when the output of the op amp is saturated low, current is being pulled through the positive feedback resistor away from the noninverting input. Thus, some of the current arriving at the noninverting input via the input resistor is being sucked away, effectively lowering the voltage at the noninverting input.

Here is the key point. When the op amp goes to high saturation because the voltage level on the noninverting input went a few millivolts above the voltage level on the inverting input, the current through the positive feedback resistor steps the voltage up a bit on the noninverting input, effectively out of the region that might result in circuit noise–induced chatter. Now the voltage on the noninverting input has to go low enough to overcome the current arriving at the noninverting input through the positive feedback resistor. When it does, the op amp's output goes to negative saturation, and the current through the positive feedback resistor causes the voltage on the noninverting input to step down, again out of the region that might result in circuit noise–induced chatter.

The overall effect of positive feedback on a comparator is to make it less prone to output chatter due to spurious noise in the circuit that may find its way to the comparator's inputs. By either increasing the resistance of the input resistor or reducing the resistance of the positive feedback resistor, we can increase the hysteresis. Conversely, if we reduce the value of the input resistor or increase the value of the feedback resistor, we will decrease the hysteresis. The hysteresis *zone* is the voltage region over which the input can vary without tripping the comparator to one of its stable saturated states.

There are special dedicated comparator ICs, and the main difference between using them and using a vanilla op amp for a comparator is speed. Op amp comparators change state from low saturation to high saturation as fast as they can, but they don't come close to dedicated comparator ICs. While op amps can go from negative saturation to positive saturation in about a microsecond or so, a dedicated comparator chip can do it in nanoseconds.

You can make your positive feedback resistor's value too low (that is, add too much hysteresis). This will cause the op amp to latch into a saturated state and stay there. The voltage applied to the input resistor feeding the noninverting input may not be high enough in amplitude to balance and overcome the current arriving there via the positive feedback resistor. If your comparator is stuck saturated high or low, increase the value of the positive feedback resistor. Another little circuit secret is to place a low-value capacitor across

Hyster-whatsis?

Typically, logic circuits and op amp comparators are designed to have *hysteresis*, or positive feedback. When an input circuit has *hysteresis*, the circuit's susceptibility to noise is greatly reduced. In effect, hysteresis widens the gap between what is considered a high input and a low input. When the input to the circuit with hysteresis goes low enough to cause a change in the circuit's output state, the input signal has to go well above that level to get the circuit's output to change.

The voltage gap over which the input has to increase or decrease to change the output state is called the *hysteresis*

zone. For example, the CD40106 CMOS Hex Inverting Schmitt Trigger has hysteresis. In order to get the output to go high, the input level must go below 1/3 of the positive supply. In order to get the output to go low the input level must go above 2/3 of the positive supply. Thus there is a 1/3 supply voltage hysteresis zone. Hysteresis helps to square up slowly rising and falling signals and reduces circuit input noise susceptibility.

the positive feedback resistor (somewhere between 5pF to 20pF) to make the op amp used as a comparator change states (go high or go low) as quickly as possible. When the output changes states, a squirt of current is pushed through the cap, giving the noninverting input a little extra jolt in the right direction.

In Figure 5-10, we see that the comparator's positive and negative excursions occur as the signal (triangle wave in the image) crosses the baseline (which is ground). If any unwanted noise finds its way into the circuit, the comparator will chatter (make several high to low transitions) as the input signal makes its zero crossings.

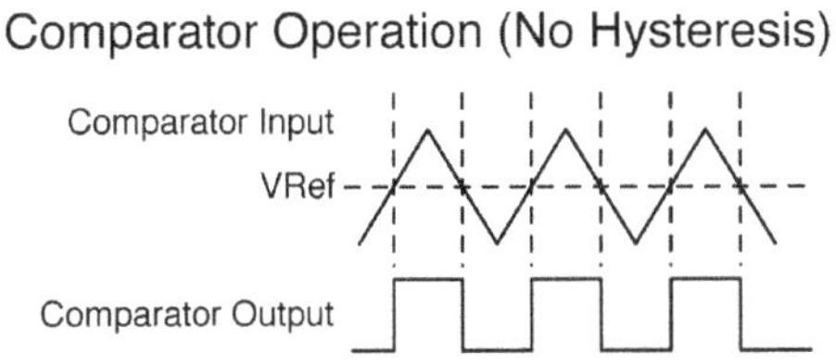

Comparator Operation (No Hysteresis)

Figure 5-10. *Comparator with no hysteresis*

In Figure 5-11, we see that the comparator's positive and negative excursions do *not* occur as the signal (triangle wave in the image) crosses the baseline (which is ground). Instead, the triangle wave fed to the comparator's input has to go somewhat above the baseline voltage before the comparator's output goes high, and then it has to go somewhat below the baseline voltage before the comparator's output goes low. The zone that the input signal has to traverse in order to make the comparator respond is referred to as the *hysteresis zone*. Since we have added hysteresis to the circuit, it will be far less prone to chatter as the triangle signal fed to the input crosses the baseline.

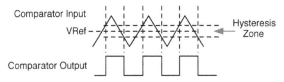

Figure 5-11. *Comparator with hysteresis*

We'll say goodbye to this section, but first I have *one more thing* to share about what specs you need to care most about when selecting op amps. The perfect op amp would have infinite input impedance, no offset voltages or currents, infinite open-loop gain, no input or output voltage limitations, and would be ultra-low power and ultra-high speed. Since we live in the real world, op amps have limitations in all of these areas. You need to read over several different op amp data sheets to familiarize yourself with their advantages and disadvantages and to find the best one for your application.

I recommend that you always use JFET input type operational amplifiers because of their high-input impedance. Select op amps that are low noise, low input offset voltage, low input offset current, and high speed. Op amps that are capable of higher speeds necessarily use more current, so if you can get by with a lower-speed op amp, do it: you'll need less supply current for it, and they tend to have better offset voltage and current specs.

Table 5-1 shows a list of some great general purpose and precision op amps widely available at the time of this writing. Over the years, you'll see that specific op amps come and go because companies are always improving on their designs and trying to come closer to the *ideal* op amp. If these particular part numbers aren't available at the time of your reading this, don't panic—just pick another general purpose JFET input (also known as BIFET input) op amp, and it'll work fine in any of the circuits shown. Always check the power supply requirements of any op amp you plan to use to make sure it will work properly with the voltage you plan to apply.

Table 5-1. Generally useful op amps

Part Number	Manufacturer	Features
TL071CP	Texas Instruments	Single op amp, JFET inputs (high impedance), low noise
TL072CP	Texas Instruments	Dual op amp, JFET inputs (high impedance), low noise
TL074CN	Texas Instruments	Quad op amp, JFET inputs (high impedance), low noise
LM324	Multiple	Quad op amp, low power
LT1097	Linear Technology	Single op amp, precision DC specs, great performer
LT1013	Linear Technology	Dual op amp, precision DC specs, great performer
LT1014	Linear Technology	Quad op amp, precision DC specs, great performer
OP275	Analog Devices	Dual op amp, precision DC specs, great performer

How the Noise Toaster Works

In this chapter, we'll go into the details of the unit's circuitry and explain the purpose of each and every component in order for you to really understand what's going on underneath the hood.

The Noise Toaster Schematics

Now let's go over each section of the Noise Toaster circuitry in detail. We'll begin our in-depth review of the Noise Toaster's circuitry with the unit's power supply.

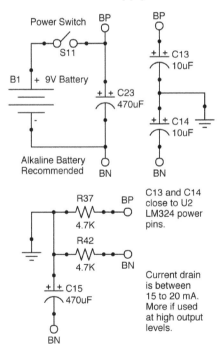

Power Supply

Figure 6-1. *Power supply schematic*

The Power Supply

In Figure 6-1, we see the unit's power supply, which is basically a 9V battery. The unit can also be powered by a clean, line-powered 9V battery eliminator. Throughout the schematics and circuit descriptions, the symbols BP and BN are used to represent the battery's positive and negative poles, respectively. Capacitor C23 (470μF electrolytic) is connected across the battery to provide a reservoir of charge to fortify the battery when peaks of current are drawn by the circuitry. When the battery starts to deplete and its internal impedance increases, C23 will be there to provide extra current when needed.

Using a Wall Wart

Some 9VDC wall warts or battery eliminators simply provide the raw, rectified output of a transformer. To work properly, they rely on a filtering capacitor in the circuit they are powering. If you use one, try to find a 9VDC wall wart that provides clean power. For additional supply filtering, you can increase the value of C23 to 1000μF (in a 16V rated cap), which should smooth out the output of just about any 9VDC wall wart.

We create a *virtual ground* for the circuit using resistors R37 and R42 (both 4.7K) and capacitor C15 (470μF electrolytic). Resistors R37 and R42 form a voltage divider that divides the unit's battery voltage in half. The positive pole of C15 connects to the junction of R37 and R42 and thus charges to half of the supply voltage. This circuit node serves as a stable *virtual ground* connection for several circuit elements. Capacitors C13 and C14 (both 10µF electrolytic) are located close to the power pins of U2 (LM324 Quad Low Power Op Amp) to stabilize its operation and provide localized reservoirs of charge for U2.

The unit draws between 15 and 20 mA, depending on whether the unit's internal amplifier and speaker or an external amplifier is used. When using the internal speaker, turning the unit's volume control all the way up will draw more current and deplete the battery faster.

The Attack Release Envelope Generator (AREG)

The Noise Toaster's AREG is a simple but effective circuit (Figure 6-2). With Manual/Repeat SPDT select switch S8 set to Manual Gate, pressing S10 (SPST N.O. PB) drops BP across resistor R61 (10K). This voltage causes current to flow to the base of Q10 (PNP transistor) via current-limiting resistor R62 (10K) biasing Q10 off. With Q10 biased off, current flows from BP through S10, current-limiting resistor R57 (200 ohms), forward-biased diode D2, Attack Time potentiometer R51 (1M), and R50 (820 ohm resistor), and charges capacitor C18 (4.7µF electrolytic capacitor). The voltage on C18 is buffered by high-impedance source follower Q9 (a 2N5457 N-Channel JFET transistor). The gate of Q9 is very high impedance, so only a very small amount of current flows through it, leaving the majority of the current to charge C18. The envelope voltage output by the AREG is delivered from the junction of load resistor R54 (4.7K) and the Source pin of Q9.

Attack Release Envelope Generator

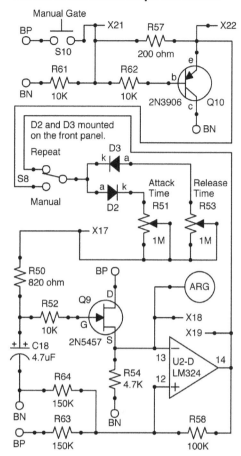

Figure 6-2. *Attack decay generator schematic*

When S10 is released, the BP voltage is no longer present across R61. Instead, BN voltage is applied to the base of Q10 via the series combination of resistors R61 and R62 (both 10K) biasing Q10 on. In this condition, current can flow between Q10's emitter and collector to BN. Q10 discharges the AREG's timing capacitor C18 via current limiting resistor R50, Release Time potentiometer R53 (1M), and forward-biased diode D3.

Thus when the Manual Gate push button S10 is pressed, the AREG outputs a voltage that rises towards BP at a rate determined by the Attack control setting. Then, on S10 release, the voltage falls toward BN at a rate determined by the Decay control setting. The Attack and Decay times are independently adjustable via pots R51 and R53, respectively, which are wired as variable resistors. The *envelope voltage* can be routed to modulate the VCO and/or the VCF via the unit's normalized switching scheme. The AREG envelope voltage is also used to modulate the unit's simple VCA.

With S8 set to Repeat, the output of the *inverting comparator* U2-D (1/4 LM324 Quad Low Power Op Amp) is connected to the junction of D2's anode and D3's cathode. Manual Gate push button S10 is not active in Repeat mode. In Repeat mode, the output of U2-D causes the AREG's output to repeatedly rise and fall in an oscillatory fashion controlled by the Attack and Decay pots.

Resistors R63 and R64 (both 150K) form a voltage divider that applies one-half of the supply voltage to U2-D's noninverting input (pin 12). However, current through positive feedback resistor R58 (100K) causes the voltage at U2-D pin 12 to be higher or lower than this, depending on the state of U2-D's output (pin 14). When U2-D's output is saturated high, its output voltage goes to about 3V above virtual ground, *pushing* current through R58 and raising the voltage at U2-D pin 12 to about 1.25V above virtual ground. When U2-D's output is saturated low, its output voltage goes to about 3.5V *below* virtual ground, *pulling* current through R58 and reducing the voltage at U2-D pin 12 to about 1.5V *below* virtual ground. Thus, the state of U2-D's output affects its threshold voltage.

When comparator U2-D's output is saturated high, capacitor C18 charges via S8, forward-biased D2, Attack Time pot R51, and current limiting resistor R50. Voltage on C18 rises toward U2-D's high saturation voltage until the output voltage of source follower Q9 dropped on R54 exceeds the threshold voltage on U2-D's noninverting input (pin 12). At that point, U2-D's output shoots rapidly to low saturation, and C18 begins to discharge via S8, forward-biased D3, Release Time pot R53, and R50. Voltage on C18 falls until the output voltage of source follower Q9 dropped on R54 goes below the threshold

voltage on U2-D's noninverting input (pin 12). At that point, the output of U2-D shoots high, and the cycle continues to repeat, resulting in oscillation. In Repeat mode, the AREG acts like an LFO, with waveform shape dependent on the Attack and Decay pot settings.

Circles with tokens in them are off-page symbols that indicate connections between the various components. Circles with the same token should be considered connected to one another.

The circled signal *ARG* (X18) gets connected to the VCF and VCO modulation switching and level controls on the front panel (see Figure 4-10 for details).

The Low-Frequency Oscillator (LFO)

The Noise Toaster's LFO provides a variety of modulation waveforms (Square, Differentiated Square, and Integrated Square). Its frequency range goes from slightly below 1 Hz to well over 200 Hz. It also has an LED rate indicator that flashes at the LFO's frequency. The schematic is shown in Figure 6-3.

The core of the LFO is a comparator, the output of which both adjusts its own threshold voltage level and provides current to charge and discharge its timing capacitor. We will assume that on power-up, the output of U2-C (1/4 LM324 Quad Low Power Op Amp) is in high saturation. Resistors R59 and R55 (both 100K) form a voltage divider that applies one-half of the supply voltage to U2-C's noninverting input (pin 10). Current through the positive feedback resistor connected between U2-C's noninverting input and its output (R60 100K) causes the voltage on U2-C pin 10 (the comparator's threshold voltage) to change, depending on whether the output is saturated high or low.

Low Frequency Oscillator

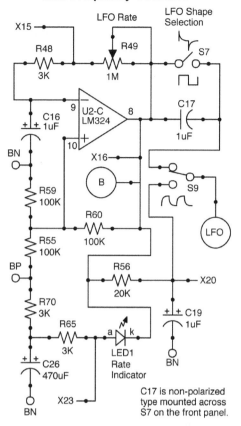

Figure 6-3. *Low-frequency oscillator schematic*

There is also a feedback path between U2-C's output and its inverting input (U2-C pin 9) via the LFO Rate pot R49 (1M) and current limiting resistor R48 (3K). The positive pole of capacitor C16 (1µF electrolytic) is connected to U2-C's inverting input. At power-up, C16's initial condition is fully discharged to the level of BN. Since the output is in high saturation (about 3V above virtual ground), current flows from U2-C's output (pin 8) through R49 and R48 and begins to charge C16 toward U2-C's output voltage. The high level of U2-C's output pushes current through the positive feedback resistor R60 and raises the threshold voltage on U2-C pin 10 to about 1V.

When the voltage on C16's positive pole rises above the voltage level on U2-C pin 10, the output of U2-C shoots to low saturation. With U2-C

in low saturation (about 3.5V below virtual ground), current is pulled through the positive feedback resistor, reducing the threshold voltage on U2-C pin 10 to about –1.25V. Capacitor C16 now begins to discharge toward U2-C's low level via R49 and R48 until it goes below the voltage level on U2-C pin 10, at which time U2-C's output shoots to positive saturation, and the cycle continues, resulting in square wave oscillation at the output of U2-C.

Resistor R70 (3K) connected to BP charges LED1's charge reservoir capacitor C26 (470µF electrolytic). When U2-C is in low saturation, current flows from C26 through current-limiting resistor R65 and forward biases LED1, causing it to glow. When U2-C is in high saturation, LED1 is reversed biased and thus off. Thus LED1 flashes at the rate of the LFO. The charge reservoir is necessary because without it, the current drawn to light LED1 causes a disturbance in the power supply, producing very slight but unwanted VCO modulation.

We modify the shape of the square wave output by U2-C in two ways. For the first waveform modification, we apply low-pass filtering to U2-C's output via resistor R56 (20K) into capacitor C19 (1µF electrolytic), which results in an integrated square wave shape. This waveform is output by the LFO when switch S9 (SPDT) is in the Integrated Square Wave position.

For the second waveform modification, we apply high-pass filtering to U2-C's output via capacitor C17. Since the capacitor blocks DC, only the rapidly changing edges of the square wave output of U2-C are passed, but the high and low levels rapidly decay, resulting in the differentiated square wave shape. This waveform is output by the LFO when S9 is in the Square/Differentiated Square position and S7 (SPST) is in the Differentiated Square position (S7 open).

The LFO's output (circuit point LFO) is directly connected to the VCO's LFO Mod Depth pot (R13 100K), but it is only connected to the VCF's Mod Depth pot (R38 100K) when VCF switch S6 is in the LFO position. Thus, the differentiated square wave's shape will be more differentiated when the VCF's Mod Source Select switch is in the LFO position, since the square wave through C17 is dropped on the parallel combination of R38 and R13 (100K in parallel with 100K = 50K).

The LFO outputs the square waveform when S9 is in the Square/Differentiated Square position and S7 (SPST) is in the Square position (S7 closed). The circled circuit point labeled LFO is the output of the LFO and is routed to the front panel controls to provide modulation for the VCO and VCF.

The LFO's Rate pot R49 adjusts the rate of the LFO from about 0.75 Hz to about 230 Hz. When the LFO is used to sync the VCO, the effect is most pronounced when the LFO's Rate knob is in the last 1/4 of its rotation.

The Voltage-Controlled Oscillator (VCO)

The Noise Toaster's VCO provides a nice, wide range of frequency (subaudible to over 15 kHz) and two waveforms (Ramp and Square). It is a ramp core design, meaning that the main *Ramp* waveform is generated by charging a capacitor with a constant current and then rapidly discharging it to zero volts when it gets to a certain amplitude. In a modular synthesizer, this basic ramp wave is often converted into several other waveforms through various electronic techniques. In the Noise Toaster's VCO, this basic ramp wave and the ramp wave converted to a square wave are provided as outputs.

The VCO is broken down into two schematics: the control voltage mixer and exponential converter, and the ramp/square wave generator.

Figure 6-4 shows the control voltage mixer and exponential converter schematic. Op amp U1-C (1/4 LM324 Low Power Quad Op Amp) is used as an inverting summer to mix the control voltages that modulate the frequency of the VCO. Three sources of frequency modulation are applied to the VCO. The VCO Frequency potentiometer R1

(100K) is used to control the initial frequency of the VCO. The ends of R1's resistive element are connected to BN and BP, and thus its wiper supplies from 4.5V below virtual ground (minimum frequency) to 4.5V above virtual ground (maximum frequency) to the inverting input of U1-C via 75K control voltage mixing resistor R2.

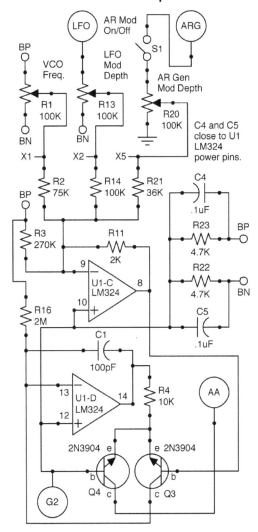

VCO - CV Mixer and Expo Converter

Figure 6-4. The VCO's control voltage mixer and exponential converter schematic

LFO Mod Depth potentiometer R13 (100K) is used to control the amount of LFO modulation applied to the VCO's control voltage (CV) summer U1-C via R14 (100K). At minimum setting, the wiper of R13 applies BN to the summer (no modulation); and at maximum setting, the full LFO output is applied to the CV summing mixer via R14 (100K).

AR Gen Mod Depth potentiometer R20 (100K) is used to control the amount of AR modulation applied to the VCO's control voltage mixer via R21 (36K) when S1 AR Mod on/off is on (closed). With R20 at minimum setting, the circuit's virtual ground is applied to the CV mixer via R21; and at maximum setting, the full AR Generator output is applied to the CV summing mixer via R21. Switch S1 (SPST mini toggle) was added because even at minimum setting, the AR generator was modulating the VCO ever so slightly.

The summing resistors used (R2, R14, and R21) are different values because the amplitudes of the modulation sources vary significantly. The summer provides more gain for modulation sources with lower amplitude. Resistor R3 (270K) biases the summing amp more toward BP to compensate for the fact that two of the VCO's modulation control pots have one end tied to BN.

I provided the VCO with its own virtual ground circuit to help prevent it from being spuriously modulated by other Noise Toaster oscillatory circuit elements. Resistors R23 and R22 (both 4.7K) divide the supply voltage in half, and the junction of the two resistors becomes the VCO's virtual ground (circuit point G2). Capacitors C4 and C5 (both .1µF) stabilize U1 and thus the VCO's operation. Circuit point G2 also appears in Figure 6-5.

The 2K feedback resistor between U1-C's output and inverting input sets its gain quite low (the gain for each input varies due to each having a different value input resistor). As an example, a 100K input resistor and a 2K feedback resistor result in a gain of 1/50th. This happens because the linear voltage to exponential current converter that follows only needs a very small change in voltage to change the frequency of the oscillator a great deal. As a matter of fact, when the output of U1-C changes by about 18mV (up or down),

the oscillator's frequency changes by approximately an octave (also up or down).

The exponential voltage to current converter is comprised of U1-D and associated components C1 (U1-D 100pF compensation cap), R4 (10K Q3 and Q4 emitter current limiter), Q3 (U1-D nonlinear feedback element – 2N3904 NPN transistor), Q4 (current sink – 2N3904 NPN transistor), and R16 (2M reference current setting resistor). Suffice it to say that the purpose of the expoconverter (as it is known) is to convert linear changes in control voltage into exponential changes in current. *Ideally*, each 18mV *increase* in the base to emitter voltage of Q3 (V_{BE}) causes Q4 to sink two times more current. For example, if Q3 at 18mV V_{BE} is causing Q4 to sink 2µA, then increasing Q3's V_{BE} to 36mV will cause Q4 to sink 4µA. Continuing to increase Q3's V_{BE} by 18mV will continue to double the current sunk by Q4. With Q3 at 54 mV, V_{BE} Q4 will sink 8µA, and with Q3 at 72 mV, V_{BE} Q4 will sink 16µA; this exponential ratio will continue to hold until Q4 is sinking a bit over 1mA.

As you may already have surmised, each *decrease* of 18mV in Q3's V_{BE} results in Q4 sinking *half* as much current. The reason I qualified this description with *ideally* is because without matched transistors and additional temperature compensation, we only get a rough approximation of this ratio. However, in the Noise Toaster, we are mainly interested in using this property to give its VCO a nice, wide frequency range, which it definitely does. The current sunk by Q4 controls the frequency of the Noise Toaster VCO's ramp generator, which will be discussed next.

For an excellent, in-depth treatise on exponential converter circuits, see Bernie Hutchins' ELECTRONOTES volume S-019, LOG AND EXPONENTIAL (ANTILOG) CIRCUITS.

Figure 6-5 shows the ramp/square wave generator schematic. As current sinks into the collector of Q4 (from Figure 6-4), it pulls current out of the inverting input of U1-A (pin 2), which is configured as an integrator. This causes the voltage on the output of U1-A to ramp up at a rate that depends on the magnitude of the current being sunk by Q4. When less current is being sunk by Q4, U1-A's output ramps up more slowly, and when more current is being sunk by Q4, U1-A's output ramps up more quickly. Thus, higher control voltage results in the VCO oscillating at a higher frequency, and lower control voltage results in the VCO oscillating at a lower frequency.

U1-B is used as a comparator whose threshold voltage is set by the resistor divider made up of R17 (10K) and R18 (15K). Since the ends of the resistor divider are connected to BP and BN, respectively, the threshold voltage applied to U1-B pin 6 is about 5.4V with respect to BN, or 0.9V above virtual ground. When the voltage level at the output of U1-A ramps to slightly above U1-B's threshold voltage, U1-B's output goes rapidly from negative saturation to positive saturation. Positive feedback resistor R7 (300K) provides U1-B with a slight bit of hysteresis so that it does not chatter when U1-A's output crosses comparator U1-B's threshold.

U1-B's output at high saturation (approximately 3V above virtual ground) drives current through forward-biased diode D1 and resistor R6 (10K), which turns on Q1 (2N5457 N-Channel JFET), allowing current to flow between its source and drain. This rapidly discharges U1-A's integrator cap C2 (.001µF ceramic), causing the output of U1-A to fall to approximately virtual ground level. Once Q1 discharges C2, U1-B's output goes to negative saturation, which reverse biases D1, allowing resistor R5 (2M to BN) to hold Q1 in the off state. This allows U1-A's output to ramp up again. This cycle repeats, resulting in a ramp waveform at the output of U1-A, whose frequency is dependent on the current being sunk by Q4. The ramp wave's amplitude is approximately 0.6V peak to peak.

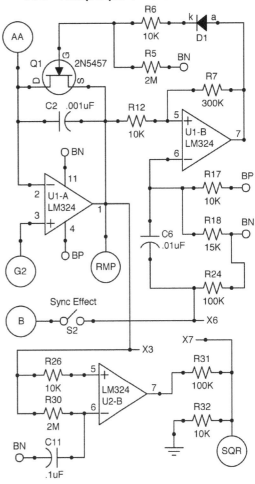

VCO - Ramp/Square Wave Generator

Figure 6-5. The VCO's ramp/square wave generator schematic

We use U2-B as a comparator to generate the VCO's square wave output. We feed the ramp wave to both the inverting and noninverting inputs of U2-B. The noninverting input sees the raw ramp wave via 10K resistor R26. The inverting input sees the average voltage of the ramp wave because we apply low-pass filtering to it by using a 2M resistor (R30) to integrate the ramp wave onto capacitor C11 (.1µF ceramic).

With these values, even at low VCO frequencies, R30 and C11 provide sufficient filtering to present the inverting input of U2-B with the average voltage of the ramp wave output of U1-A.

When the amplitude of the ramp wave applied to U2-B's noninverting input (pin 5) is above the average ramp wave voltage applied to U2-B's inverting input, the output of U2-B shoots to positive saturation. When the amplitude of the ramp wave applied to U2-B's noninverting input is below the average ramp wave voltage applied to U2-B's inverting input, the output of U2-B shoots to negative saturation. Thus, a square wave is present at the output of U2-B, the frequency of which is the same as the ramp wave's.

Resistors R31 (100K) and R32 (10K) comprise a voltage divider that reduces the amplitude of the square wave prior to applying it to the input of the VCF to about 550mV peak to peak.

When switch S2 is turned on, we apply the square output of the low-frequency oscillator (LFO) to the junction of C6 (.01µF ceramic) and R24 (100K). A narrow negative-going pulse is conducted to the threshold bias of U1-B via C6 when the LFO's square wave goes low. This changes the threshold voltage enough to cause the integrator to reset, which results in an unusual waveform at the output of the ramp generator that contains harmonics of both the LFO and the VCO. The effect is most pronounced when the LFO is in the last 1/4 of its adjustment range and the VCO's frequency is swept from low to high.

The White Noise Generator

When you exceed the emitter to base breakdown voltage (BV_{EBO}) of an NPN transistor, it behaves like a zener diode. A side effect of the zener breakdown mechanism is the generation of *white noise*. The white noise signal created in this manner is very low amplitude and must be amplified in order to be used in a synthesizer. We couple the low-amplitude white noise generated at the junction of current limiting resistor R8 (1M) and the emitter of Q5 (2N3904 NPN transistor) to a two-stage NPN transistor amplifier to boost the signal. A schematic is shown in Figure 6-6.

Capacitor C3 (.1µF ceramic) couples the low-level noise signal to the base of Q2 (2N3904 NPN transistor), which is configured as a common emitter

amplifier. Collector resistor R9 (10K) and collector-to-base resistor R10 (470K) bias Q2 into the operating region that permits the noise signal to be amplified with minimal distortion. Negative feedback provided by R10 also helps stabilize Q2's quiescent operating point (Q). The amplified signal at Q2's collector is capacitively coupled to the two-transistor amplifier's second stage via capacitor C7 (.1µF ceramic). The second stage's gain is reduced to avoid clipping by the inclusion of R25 (100K) between the coupling cap (C7) and the base of Q6 (2N3904 NPN transistor). The second stage is biased in the same manner as the first by collector resistor R15 (10K) and collector-to-base resistor R19 (470K).

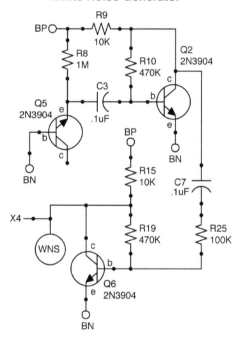

White Noise Generator

Figure 6-6. White noise generator schematic

The first stage provides a gain factor of roughly 250 and the second stage a gain factor of roughly 3. The total gain factor of the amplifier (750) boosts the gain of the low-level white noise to roughly 1.5V peak to peak at circuit point WNS (X4). The white noise generator output level may be higher or lower depending on the amount of white noise generated by Q5, which should be selected for best noise. The white noise is applied to the VCF's input via VCF Input Selection switch S4 (SPST). When the unit's battery becomes too depleted, the white noise generator will stop because the supply voltage won't be high enough to cause Q5's EB junction to enter zener breakdown and produce noise.

The Voltage-Controlled Low-Pass Filter (VCF)

The VCF in the Noise Toaster adds a lot of sonic character to both the VCO output and the white noise generator. The VCF is an active, low-pass "T" type filter (due to the two caps and resistor to ground in the feedback loop resembling a "T") built around U2-A (1/4 LM324 Low Power Quad Op Amp). A schematic is shown in Figure 6-7. Feedback resistor R35 (4.7M) and the two filter capacitors C9 and C10 (both .001µF ceramic) determine the filter's characteristics and gain. The filter's cutoff frequency increases as we decrease the resistance between the junction of the two feedback capacitors and virtual ground by adjusting Cutoff Frequency potentiometer R28 (100K). The filter's Q increases as we decrease the resistance between the two feedback caps of the filter by adjusting Resonance potentiometer R29 (100K).

In order to permit the filter's cutoff frequency to be voltage-controlled, we use Q8 (2N5457 N-Channel JFET) in its voltage-controlled resistor (VCR) mode. As we apply more positive voltage to the gate of Q8, the resistance between its drain and source goes down, resulting in an increase of the filter's cutoff frequency. We bias the voltage that is fed to Q8's gate so that its drain to source resistance varies nicely as we apply the voltages from the unit's modulators (LFO and AREG).

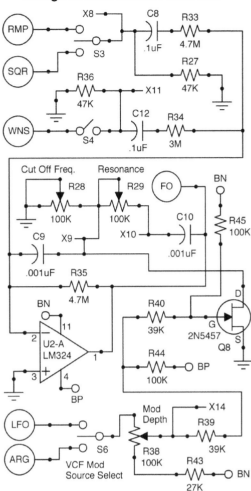

Voltage-Controlled Low-Pass Filter

Figure 6-7. *Voltage-controlled low-pass filter schematic*

Switch S6 (SPDT C.O.) selects the modulation source for the VCF. It can be either the LFO, the AREG, or none (center off position). The modulation source selected by S6 is dropped across Mod Depth potentiometer R38 (100K), which is wired in series with R43 (27K) to BN. The pot's resistive element and R43 form a voltage divider that causes the minimum modulation voltage to be above the level of BN. Without R43, the Mod Depth pot does not start affecting the VCF's cutoff frequency until the second half of its rotation. Because JFETs vary, you may need to increase or decrease R43's value to maximize the range of the Mod Depth pot.

Resistors R39 (39K), R40 (39K), R44 (100K), and R45 (100K) both current-limit and bias (to both BP and BN) the modulation sources to permit Q8 to work in its resistive region. When the modulation signals are increased via adjustment of Mod Depth pot R38, the transistor functions as a voltage-controlled resistor and serves the same purpose as the cutoff frequency pot (i.e., bringing the junction of C9 and C10 closer or further from ground). The filter rings slightly, even when the Resonance potentiometer R29 (100K) is at its minimum setting, but rings quite a bit more when the Resonance control is turned up.

The inputs to the VCF come from the VCO and the white noise generator. Switch S3 (SPDT C.O.) selects filter input from among the following: VCO Ramp wave, VCO Square wave, or none (center off position). Switch S4 (SPST) applies white noise to the filter's input when closed.

The input signals are capacitively coupled to the filter's input to avoid DC offset issues. Switch S3's center pole drops the selected VCO waveform across R27 (47K), where it is capacitively coupled to the input of the VCF via mixing resistor R33 (4.7M). Switch S4, when closed, drops the white noise signal across R36 (47K), where it is capacitively coupled to the input of the VCF via mixing resistor R34 (3M). U2-A pin 3 is connected directly to the unit's virtual ground. Resistors R27 and R36 (both 47K) stabilize the inputs to the filter when the switches that apply the VCO waveform and/or white noise to its input are open. With the VCO's Square wave applied to the filter, Resonance at maximum, and Cutoff Frequency adjusted to be resonant at the VCO's frequency, the VCF's output amplitude is approximately 4V peak to peak, oscillating about virtual ground. The VCF's output is taken from circuit point *FO* and is connected directly to the input of the VCA.

The Voltage-Controlled Amplifier (VCA)

The Noise Toaster's simple VCA is lo-fi in that it distorts the signal going through it a bit (flattening the wave a bit on one side), but it still passes the signal faithfully enough to provide some interesting effects. The VCA is comprised of Q7 (2N5457 N-Channel JFET transistor) used in VCR (voltage-controlled resistor) mode and its associated biasing resistors. A schematic is shown in Figure 6-8.

Voltage-Controlled Amplifier

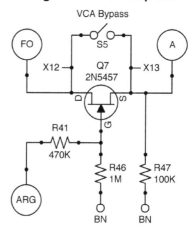

Figure 6-8. *Voltage-controlled amplifier schematic*

The output of the VCF (circuit point FO) is applied directly to the VCA's input, which is the Drain pin of Q7. The output of the VCA is Q7's Source pin, which goes to BN through load resistor R47 (100K). Resistors R41 (470K) and R46 (1M) bias Q7's gate so that the output of the AREG modulates the amplitude of the signal passing from Drain to Source most effectively. As voltage applied to circuit point ARG is made more positive by the output of the AREG, the resistance between the Drain and Source goes down, thus dropping the signal applied to Q7's Drain onto R47. Q7's Source (the VCA's output) is applied to the unit's Volume pot R66 (100K). The VCA's modulation input (circuit point ARG) is hard-wired to the output of the AREG and thus only the AREG's settings affect the VCA's output amplitude when

S5 (SPST) is open (VCA (AR mod) position). When S5 is closed (Bypass position), the VCA is bypassed entirely.

The Audio Amplifier

The Noise Toaster's audio speaker is driven by U3, an LM386N-4 1W integrated audio amplifier (the schematic is shown in Figure 6-9). We add the components that the LM386N data sheet tells us to: capacitor C24 (.047µF ceramic), R69 (10 ohm), C22 (220µF electrolytic), and an 8 ohm speaker and voila—we have a complete 1W amplifier. I also added the recommended bypass cap C25 (10µF electrolytic) and C20 (.1µF) across the chip's power pins for good measure. To read complete information about the operation and internal circuitry of the LM386N, search for it online and read its data sheet.

The output of the simple JFET VCA is fed to one end of volume pot R66, the other side of which is connected to BN. We block DC from getting to the input of U3 via C21 (1µF bipolar nonpolarized aluminum). As we advance R66, we pick off more and more VCA output—and if we go the other way, less and less. Resistors R67 (100K) and R68 (20K) further attenuate the level fed to the amplifier's input from the volume pot so that we don't overdrive the amplifier, resulting in clipping and distortion.

The line out jack's tip terminal is fed from AC coupling cap C21 (1µF bipolar aluminum), which connects to the 100K volume pot's (R66) wiper. The input to the LM386 amplifier (point X27) connects to the switch leg of the 1/4" jack. When no phone jack is inserted into J1, the jack's switch leg contacts the tip tong and routes the signal coming through C21 to the internal amplifier. When a jack is inserted into J1, the tip tong is forced away from the switch leg, and the signal is no longer fed to the internal amplifier's input. The unit draws the lowest current (approximately 15mA) when using line out. Turning the unit's volume up when using the internal amplifier can drain up to 40mA or more from the battery, reducing its lifespan significantly.

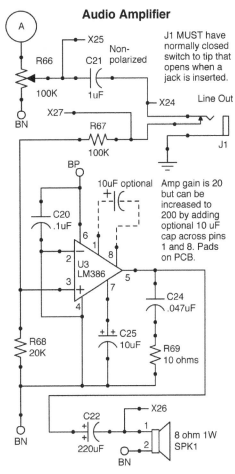

Figure 6-9. Audio amplifier schematic

Your Electronic Music Studio   7

When the last wire is soldered and the last screw is tightened and the analog synthesizer project you just built with your own hands is sitting there in front of you, it's time to put it to work making some cool sounds. Give your shop a cleaning, clear the papers from your desk so you can see the top of it again, get a cup of tea, cocoa, coffee…whatever you like to drink, and get ready for some creative fun. To me, recording the sounds you create with your synthesizer and making some creative contribution to the human experience is the ultimate goal.

In this chapter, I plan to introduce you to some tips, techniques, and creativity-boosting exercises you can use with your multitrack software to produce some cool-sounding recordings with the gear you build. With multitrack recording and the effects provided by most digital audio workstations (DAW), you can make your Noise Toaster or other project sound as impressive as any analog synth out there. Recording exercises I'll cover will include the following:

Musique Concrete

> *Musique Concrete* is a technique for producing experimental musical compositions by splicing together in unexpected and interesting ways pieces of audio tape that contain interesting sounds together. Using a DAW enables us to use WAV file snippets instead of actual magnetic tape, which is *way* more convenient.

Multitrack Layering

> By building up layers of sound, you can make even a simple analog synthesizer sound like a B52 going over.

Multitrack Volume Envelope Layering

> Using volume envelopes generated by the recording software, various tracks can fade in and fade out, providing interesting audio effects.

The DAW we will be using for illustrative purposes in this chapter is version 2.0.2 of Audacity recording and editing software. You can download it from *http://audacity.sourceforge.net/*. It is free software distributed under the terms of the GNU General Public License; if you find that it provides you with useful functionality, make a donation to the Audacity project to help keep new and better features coming.

Connecting a Project to Your Computer's Sound Card

To use the computer to record the sounds of your project, you must be able to connect the output of your project to it. To get you started, I'll present a simple circuit (see Figure 7-1) that you can use to connect your synth to your computer sound card's *line* input. We use the line input because the amplitude of most analog synth outputs is too high and will overdrive the sound card's microphone preamp, resulting in distortion. The simple circuit presented takes the line-level output of your synth and allows you to apply it to both stereo inputs equally or pan the mono signal within the stereo field.

Figure 7-1. *Sound card audio interface box*

Of course, if you already have an audio mixer, you just need to get an adapter cord that goes from the line outputs of your mixer to the line input of the sound card. The output of your mixer may be 1/4" jacks or RCA-type jacks, but the input to the majority of computer sound cards is a stereo 3.5mm (1/8") minijack. Connect the mixer's output to the sound card's line input and you're ready to go. You can skip the next section and go straight to the creativity exercises.

If you've got a laptop, replace the term "sound card" in this discussion with the phrase "audio input on your laptop."

The schematic for the simple mono to stereo audio interface is presented in Figure 7-2. The input jack's tip connector is connected to R1 and R2 (both 2K 5% 1/4W resistors), the other ends of which are soldered to the two end terminals of pan pot R4 (10K linear pot). Two more resistors R3 and R5 (also both 2K 5% 1/4W) connect from the pot's end terminals to the L and R stereo output jack tip terminals. The ground from both jacks get connected together and to pot R4's wiper terminal and body.

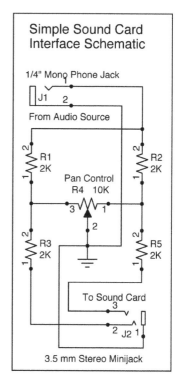

Figure 7-2. *Sound card audio interface schematic*

With the wiper of R4 centered, the input signal splits to flow through R1 and R2, where each path is dropped across half of pot R4's resistive element, since its centered wiper goes to ground. The signals dropped on R4 continue through resistors R3 and R5 to the two stereo output jack terminals. This jack is connected to your computer sound card's *line* input by means of a cable with 3.5mm stereo connectors on both ends. With R4 centered, the signal splits evenly to both the left and right channels. However, if you change R4 from its centered position, you bring either the junction of R1, R3 or the junction of R2, R5 closer to ground and thus attenuate the signal in that path more than the other. By adjusting pot R4, you can *pan* the signal in the stereo field from left to right in real time while recording.

3.5mm versus 1/8" jack. From what I could gather, jacks and plugs of either size are compatible. The diameter of the hole in a 1/8" jack I bought from Radio Shack measured 0.14", which is slightly above 3.5mm.

Table 7-1 contains the parts list for the simple audio interface.

Table 7-1. Simple audio interface bill of materials

Qty	Description	Value	Designators
1	Linear Taper Potentiometer	10K	R4
4	Resistor 1/4 Watt 5%	2K	R1, R2, R3, R5
1	3.5 mm Stereo Minijack	RS-Catalog#: 274 3.5 mm Stereo Minijack	J2
1	1/4" Mono Jack	RS-Catalog#: 274 1/4" Mono Phone Jack	J1
1	Project Enclosure (4x2x1")	RS-Catalog#: 270-1802	J1

Figure 7-3 shows how the simple interface box is wired. Drill a 5/16" hole for the pot, a 3/8" hole for the jack, and a 1/4" hole for the 3.5mm (1/8") jack. Mount the components and then solder the resistors between the jacks and pot terminals as shown. After installing the resistors, wire together the jack ground terminals and to pot R4's wiper terminal and metal body. I used an emory board to roughen the pot body's surface, a bit higher heat, and some additional flux to get the solder to flow properly.

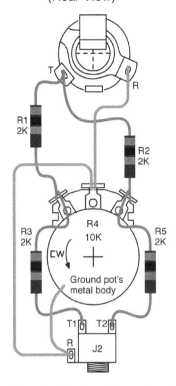

Simple Sound Card Interface Wiring (Rear View)

Figure 7-3. Sound card audio interface wiring (I)

Figure 7-4 presents another view of the project box I used showing the wiring and pot grounding. Since I used a plastic box, I had to connect the grounds of the jacks to the R4's body as well as its wiper. All jacks are not created equal, so be sure of which terminal on your jack is the tip and which is the ground. Ring it out with your multimeter to be sure.

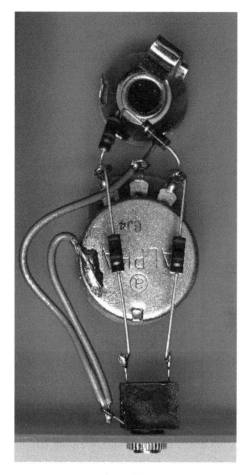

Figure 7-4. Sound card audio interface wiring (II)

Figure 7-5 shows how the simple audio interface connects your synth to your computer's sound card and ultimately to your DAW software.

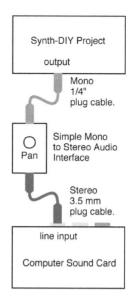

Figure 7-5. Sound card audio interface block diagram

Connect your synth's output jack to the mono input jack of the audio interface box, and connect the audio interface box's stereo output jack to the line input of your sound card (usually the light blue colored jack) using a cable with 3.5mm stereo plugs on both ends (Figure 7-6).

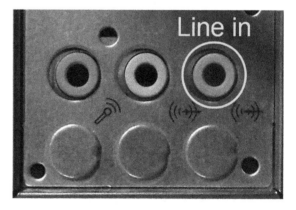

Figure 7-6. Computer sound card inputs highlighting line input jack

Once you have audio into the sound card's line input, you're ready to start using your recording software or DAW.

Introducing Audacity

Audacity is free audio recording and editing software developed by a group of volunteers and distributed under the GNU General Public License (GPL). Audacity is open source software; its source code is available for study and/or use. The Audacity team welcomes donations to support Audacity development. Individuals can also support the project by donating time and effort to help with documentation, translations, user support, and code testing.

This section is *not* going to be a tutorial on how to use Audacity because the Audacity folks have

already done an outstanding job with that. Online help for Audacity can be found at *http://audacity.sourceforge.net/help/*. We are going to cover various recording techniques that can be carried out with any DAW, using Audacity to illustrate the examples.

The Audacity Interface

In Figure 7-7, we see the Audacity interface. Although this will not be a full tutorial of how to use Audacity, I'll cover enough of its operation to get you started.

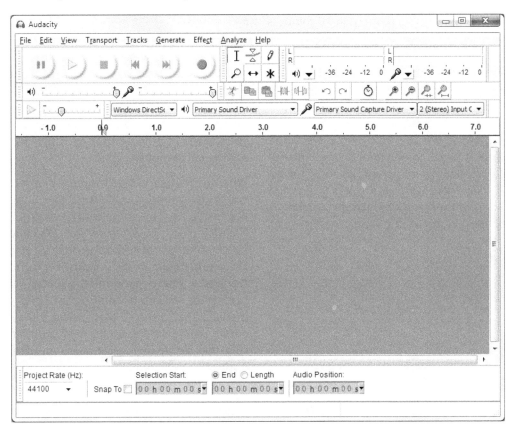

Figure 7-7. *Audacity recording software interface*

Sound Card Interface Discovery

Audacity should find your computer's sound card or alternate audio interface and allow you to choose both the audio input source and the

audio output source. Consult the online Audacity help if you encounter any issues related to your computer's audio interface. Audacity works under Windows, Mac OS X, and all flavors of Linux.

Saving Projects

To save all of the tracks and settings for a project choose the *File→Save Project* menu item. It's always a good idea to save any audio project at various stages using a letter or number filename suffix to insulate yourself from unrecoverable mistakes or worse, an undocumented software feature (more commonly known as a bug). No offense to the great people writing DAW software, but as every software engineer knows, *no* software is bug-free. You might save your work every 10 minutes or so using the following file naming sequence: MyProjectName_001.aup, MyProjectName_002.aup, etc. Doing so will allow you to go back to a previous version and prevent you from losing more than 10 minutes or so of work.

Exporting Audio Formats

When you have created one or more tracks of audio data and are ready to save your composition in a format that is compatible with popular media players, do the following. Select the *View→Mixer Board…* menu item to open the Audacity Mixer Board window (Figure 7-8). The Audacity Mixer Board window displays a group of slide faders (one per track), which are used to *mix* (adjust the volumes of) the various tracks.

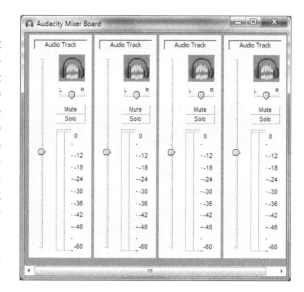

Figure 7-8. *Audacity Mixer Board window*

The faders are active during playback and provide real-time adjustment of each track's volume and pan position. Each track's Mute and Solo buttons are also active during playback, allowing you to silence (mute) or listen exclusively to (solo) a track with a mouse click.

When you have adjusted the *mix* exactly the way you like it, it's time to *export* the audio data. Select the *File→Export* menu item to open the Export File dialog box. Audacity version 2.0.2 and later supports export of audio data into an extensive list of formats (Figure 7-9). Choose the format and storage location desired and then click *Save*. Play back your exported composition using your media player to ensure that everything went according to plan. If you encounter a problem, consult the extensive online help available for Audacity.

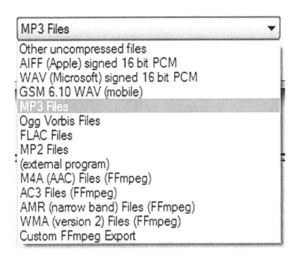

Figure 7-9. Audacity audio export formats

Transport Controls

The large transport controls (from left to right) are used to *pause* playback or recording, *play* previously recorded material, *stop* recording or playing, *move* to the beginning of the recorded material, and *move* to the end of the recorded material, respectively (Figure 7-10).

Figure 7-10. Audacity transport controls

Clicking the Record button creates a new track and simultaneously begins recording on it. If you want to record on the currently selected track, hold the Shift key while clicking the Record transport button. If the selected track contains previously recorded material, recording will begin at the end of the track's current material. If the track is empty, recording will begin at the selected position within the track.

Remember: holding Shift and clicking Record records on the selected track. Clicking Record (without Shift) creates a new track and begins recording from time zero.

Audacity Tools

The tools in Figure 7-11 can be chosen one at a time, or you can use the context-driven *Multi-Tool Mode* (asterisk icon), which will automatically select the appropriate tool for the current task. Experiment with this mode to learn its subtleties, if you elect to use it. The Audacity Tools are modal in nature (except for the contextual *Multi-Tool Mode*) and determine how mouse actions manipulate the audio data within a track.

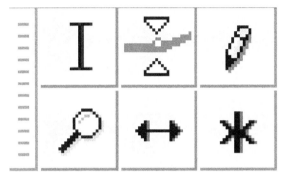

Figure 7-11. Audacity Tools toolbar

Starting at the upper left, we see the *Selection Tool*, which is used to either select a position within a track (click) or select a section of recorded material (click and drag). The selected area can be cut and pasted to another location on the same or a separate track. To the right, we have the *Envelope Tool*, which is used to apply a volume envelope to previously recorded material. To the right of the *Envelope Tool*, we see the *Draw Tool*, which is used to edit individual audio samples. We will not be using the *Draw Tool* in our experiments.

At the lower left, we see the *Zoom Tool*, which is used to horizontally expand or contract the track audio-data view. Zoom-in (horizontal expansion) or zoom-out (horizontal compression) is applied to the track audio data by clicking the left or right mouse button, respectively. Zooming can also be accomplished by using Ctrl-Mouse Wheel. Holding the Ctrl key while rolling the mouse wheel away zooms in and rolling the wheel in the other direction zooms out.

Next is the *Time Shift Tool*, used to drag selected sections of previously recorded material to the left or right. Finally, the *Multi-Tool Mode* button lets Audacity determine which tool to enable based on the context of the current operation.

Zooming can also be accomplished using the convenient zoom toolbar (Figure 7-12), which contains, from left to right: Zoom In, Zoom Out, Fit Selection, and Fit Project. Fit Selection zooms in on the selected region, and Fit Project zooms out to allow you to see the full horizontal scope of the project.

Figure 7-12. *Audacity zoom tools*

Audacity Edit Toolbar

The edit toolbar (Figure 7-13) contains the usual suspects (Cut, Copy, and Paste) and two additional useful tools: Trim and Silence. When *Trim* is clicked, all nonselected audio data is deleted (or trimmed), while the selected region is left as it was. When *Silence* is clicked, all selected audio data is removed and replaced with silence.

Figure 7-13. *Audacity edit tools*

Audacity has several ways to delete material from a track. Select the audio clip to remove, and then click one of the following key combinations to remove the material as follows:

Ctrl-K
> Deletes selected material. If the selection is within the boundaries of a clip, the two remaining end sections are joined together.

Ctrl-Alt-K
> Deletes selected material. If the selection is within the boundaries of a clip, the clip is split into two clips, and the middle section is removed. The two remaining end sections are not moved or joined.

Sound-on-Sound Recording

Out of the box, Audacity is set to play back previous tracks while recording new ones (often referred to as sound-on-sound recording), which is generally what you want. You can change this by means of the *Edit→Preferences→Recording* menu path, which opens the *Preferences : Recording* dialog. Alternatively, each track has a *Mute* control, which allows you to silence it during playback or recording of other tracks. Additionally, each track has a *Solo* button, which permits you to hear just that track when playing back material (Figure 7-14).

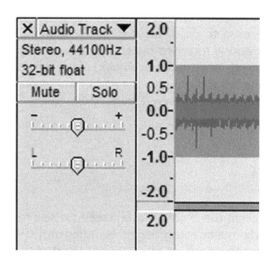

Figure 7-14. *Audacity track mute and solo controls*

Audacity Effects

Audacity comes with a plethora of cool effects that you can apply to your audio material to enliven it and imbue it with ambience, space, depth, and interest. You can find excellent online help as well as videos made by users of Audacity regarding any effect of interest by searching online.

Figures 7-15 and 7-16 show the list of effects that come with version 2.0.2 of Audacity. We will only be using a handful of the effects listed in the audio experiments to follow. If an effect name is followed by an ellipsis (*Amplify...*, for example), it indicates that the effect has settable parameters. When you select an effect ending with an ellipsis, a dialog box will open displaying controls that are used to adjust the effect's parameters. After the parameters have been set as desired, clicking the OK button will apply the effect. Effect names not ending in an ellipsis are applied to the selected material as soon as you click the effect name (*Fade In*, for example).

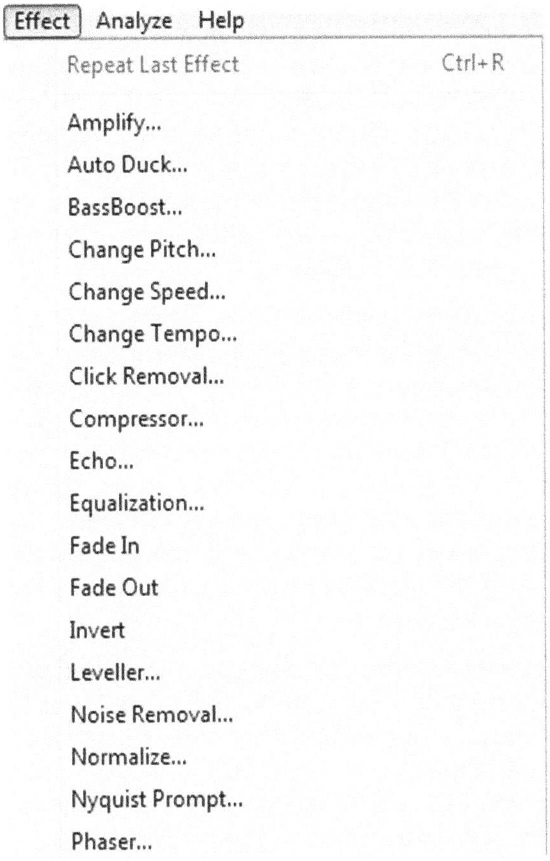

Effect	Analyze	Help

Repeat Last Effect	Ctrl+R
Amplify...	
Auto Duck...	
BassBoost...	
Change Pitch...	
Change Speed...	
Change Tempo...	
Click Removal...	
Compressor...	
Echo...	
Equalization...	
Fade In	
Fade Out	
Invert	
Leveller...	
Noise Removal...	
Normalize...	
Nyquist Prompt...	
Phaser...	

Figure 7-15. Audacity Effect List I

This is the remainder of the list of effects available in Audacity.

Repair
Repeat...
Reverse
Sliding Time Scale/Pitch Shift...
Truncate Silence...
Wahwah...

Clip Fix...
Cross Fade In
Cross Fade Out
Delay...
GVerb...
Hard Limiter...
High Pass Filter...
Low Pass Filter...
Notch Filter...
SC4...
Tremolo...
Vocal Remover (for center-panned vocals)...
Vocoder...

Figure 7-16. Audacity Effect List II

Effect names suffixed with an ellipsis have an associated dialog with settable parameters. Effect names with no ellipsis are applied upon selection of the effect.

Some effects work between the two stereo tracks and require the tracks to be split into mono or reunited as a stereo track to accomplish application of the effect. The Vocoder is an effect of this type. Figure 7-17 shows the track drop-down menu where you find the operations used to accomplish these tasks.

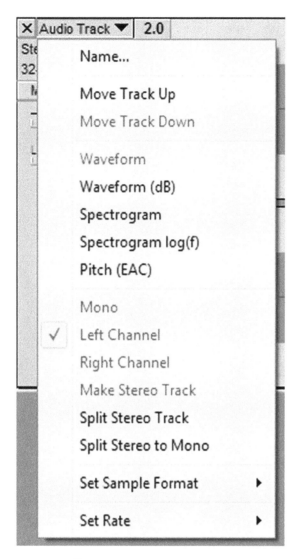

is affected than you intended, simply use the excellent *undo* (Ctrl-Z) functionality designed into the application.

> *Remember: effects are applied to all selected clips. If no clips are selected, effects are applied to all clips in all tracks.*

Effect Tails

Often, to produce a sonic effect, it is important to include what is known as an *effect tail* (see Figures 7-18 and 7-19). For example, if you apply reverb to an audio clip, the reverberation effect is heard as the clip plays, but when the clip ends, there is no trailing reverberation (effect tail) produced. The same thing applies to a delay or echo effect applied to a track. Without the effect tail, the sound effect falls flat.

In order to achieve an *effect tail* for a clip, you need to add a few seconds of silence after the section you want the effect to tail on. To add a section of silence to any clip, select an empty part of the track and then navigate to the menu item *Generate→Silence….* This will open the Silence Generator dialog box, where you can specify the duration of the silence. Upon opening, the Silence Generator dialog displays the duration of the selected section.

Once the section of silence has been created, you must join it to the clip for which you want to generate the effect tail. For example, in this clip I would like to add reverb only to the last sound burst. To do so, I first add a section of silence to the right of the clip, as shown in Figure 7-18.

Figure 7-17. *Track drop-down menu*

When applying an effect, it is important to note that it will be applied to *selected clips* or, in the case where no clips are selected, to *all* clips in *all* tracks. If you apply an effect and more material

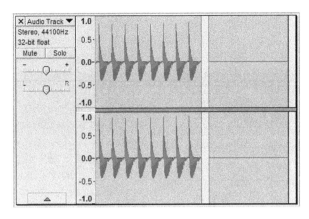

Figure 7-18. *Creating an effect tail (I)*

Next we join the silence clip to the preceding clip by selecting both clips and choosing *Edit→Join* from the menu. Once the section of silence is joined to the other section, we split the last sound burst and silence section away from the whole by first selecting it and then choosing *Edit→Split* from the menu. Once separated, select the section that contains the last sound burst and

following silence and apply an effect. For our example, we applied the *GVerb…* effect. Once the effect is added to the separated section, we rejoin it to the preceding section. Figure 7-19 displays the resulting clip. The last audio burst alone has the reverb effect, which tails off into the added silence section.

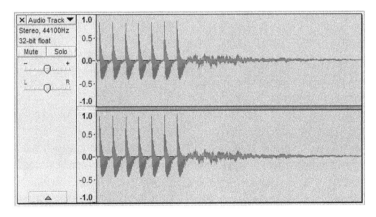

Figure 7-19. *Creating an effect tail (II)*

As previously mentioned, Audacity has a *lot* more features and capabilities than I've presented here, and I encourage you to explore the program's online help once you've mastered the basics and need more advanced features.

In the following recording exercises, effects can be added to any track or clip to add dimension and depth. Adding effect tails, where appropri-

ate, adds ambience and depth to your sonic compositions.

Recording Exercises

There are times when the muse of creativity refuses to sit on your shoulder and cooperate. This is perfectly normal; everyone has more or less creativity at times. Maybe it's bio-rhythms, the alignment of the stars, whether you took your vitamins…who knows? However, you can always keep your *technical* skills sharp, whether that means doing scales on your instrument, playing and producing songs by other artists you enjoy, using MIDI files to control your synthesizer, or engaging in what I refer to as *creativity exercises* with your analog synth and DAW.

Having a particular form to work within may actually help to ignite your creativity. You start by following the steps in the exercise and then, more often than not, you find yourself saying… *"Hey you know what else would sound cool…"* and Bob's your uncle, creative juices are flowing again. It won't hurt a thing to give these exercises a try to see what you can come up with. And don't forget, if you come up with something totally cool, post it to *The Art of Electronic Sound* project (*http://bit.ly/14M5tQ2*) on the MFOS website for other electronic builder/musicians to check out. Submitting works to this project is free. *Yours may be the work discovered by a famous filmmaker looking for an original score.* Let's take a closer look at these creativity exercises.

Musique Concrete

The idea of recording sounds on magnetic tape and then manipulating them has been around since the development of the reel-to-reel tape recorder. The term *Musique Concrete* was reportedly coined by Pierre Schaeffer, who was an audio engineer and early experimental music pioneer working at a radio station in Paris in the 1940s. I find it interesting that imaginative people have been finding ways to make new and interesting sounds using electronics since it was technically possible to do so.

In Musique Concrete, the *composer* first records various natural sounds, electronic sounds, and/or speech onto magnetic tape. Next, she cuts sections of the tape so that each section contains a sound of interest. Now the composer can manipulate each tape section by hand, sliding a section forward or backward over a playback head and recording the result onto a second tape recorder, for example. Another popular Musique Concrete technique is to splice the individual pieces of sound containing magnetic tape in an unusual order to make a conglomeration of experimental sound.

To get started with this experiment, connect the output of your synth to your computer's sound card, as shown in the previous section. Adjust your synth for an interesting sound and go ahead and open a track by clicking the Record transport control. Record a second or so of the sound.

Adjust the unit for a contrasting sound. Contrasting sounds, for example, are high and low pitch, high and low filter cutoff frequency, and high frequency and amplitude VCO modulation versus low frequency and amplitude VCO modulation. Experiment to find a variety of sounds and record snippets of them as shown in Figure 7-20. Remember, you can include natural sounds or voice as well.

Repeat this procedure several times, adjusting your synth for an interesting or contrasting sound and then using the Shift key and Record to append the newly recorded material to the same track. When you have added four or five interesting and contrasting sounds, you are ready for the next step. Figure 7-20 shows the synth snippets I recorded.

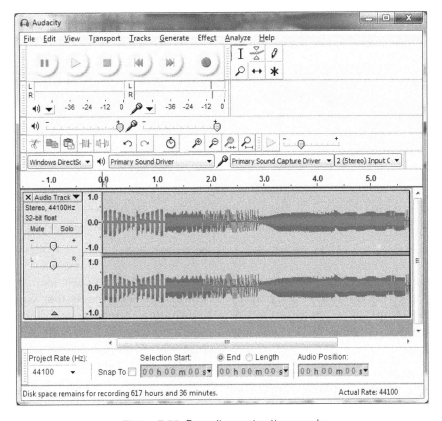

Figure 7-20. *Recording contrasting sounds*

Select the *Time Shift Tool* and drag the recorded material to the right to open some space on the left side (beginning) of the track. Now, using the *Selection Tool*, select small pieces of the recorded material and paste them near the beginning of the track next to one another. Use the *Time Shift Tool* to drag the small clips you're placing on the left side of the track so that they abut one another as shown in Figure 7-21.

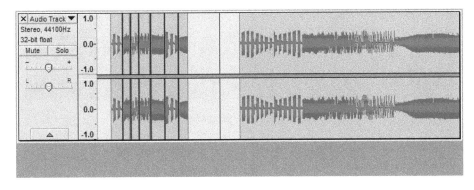

Figure 7-21. *Selected snippets prior to clip join*

Once you have the selected clips dragged together, select them all with the *Selection Tool* and then invoke the *Edit→Clip Boundaries→Join* menu item, which will cause the group of clips to become one contiguous clip. Figure 7-22 shows the results of joining the clips.

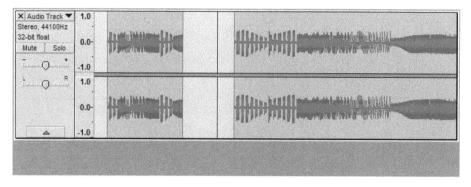

Figure 7-22. *Selected snippets after clip join*

Clicking a position on the time line (numbered line just above the track window) begins playing nonmuted tracks from that point. Listen to the joined snippets by clicking on the time line just to the left of the clip.

Delete the originally recorded material from which we cut the snippets by selecting it and pressing Ctrl-K. Now copy and paste the newly created clip conglomeration into the track a few times and listen to it. With practice, you will be able to build up some interesting sound clips in this manner (one example is shown in Figure 7-23). Remember that you can also include natural sounds by recording with the computer's microphone into Audacity. Audacity lets you apply effects to the clips that are very similar to those applied by the original Musique Concrete pioneers who used tape recorders. You can speed up or slow down a clip (*Effect→Change Speed...*), play a clip backwards (*Effect→Reverse*), or change the pitch of a clip (*Effect→Change Pitch...*).

Figure 7-23. *Our first attempt at Musique Concrete*

Experiment with Musique Concrete for a while. Now that you've got the idea, try producing sounds that would fit these scenarios:

- An excited R2-D2
- Factory noises
- A robot walking
- A robot talking
- A spaceship
- A flying saucer

To produce an interesting audio effect, prior to joining the separate copies of the clip conglomeration, apply different effects to each of them to see how it can add interest and contrast to your first Musique Concrete composition. Select each clip in turn, and then select an effect you think would be cool and apply it. If you don't like the result of applying an effect, simply press Ctrl-Z (undo) to go back to the pre-effect state.

I hope this introduction to Musique Concrete has been informative and that you come back to it from time to time to experiment and perhaps find your muse. Another engaging sonic experiment consists of building layers of sound to create unique and exotic soundscapes. This is called multitrack layering, and we'll examine it next.

Multitrack Layering

No matter what type of synthesizer you own, be it the simplest noise box or the much revered ARP-2600, multitrack layering can help you achieve incredibly textured soundscapes. This section will cover creating multiple audio tracks, each of which will contain an element of the total sound. By using the real-time Audacity Mixer Board, you can experiment by subtly changing the contribution of each track to the total composition.

Begin by adjusting your synthesizer to create the first layer of sound. For this demonstration, I'll start by creating a track containing the sound of a howling wind. To make the howling wind sound, I pass white noise through a low-pass filter with a bit of resonance and slowly adjust the filter's cutoff frequency up and down. When you're ready, click the Record transport control to create a new track and begin recording. Record the track for as long as you want the entire composition to be. Save your project.

Next, adjust your synth's VCO to produce a higher frequency square wave (between 2 kHz and 3 kHz) and route it through the VCF and VCA. Adjust the VCF's cutoff frequency to pass most of the signal. Control the VCA with an AREG set for fast attack, slight decay, and manual gating. Now begin recording a new track by clicking the Record transport control. While recording, manually gate the AREG randomly with intervals of a second or so between gates. Before each gate is applied, manually change the VCO's frequency slightly using its Frequency control. The goal is to have a variety of high-pitched *dinks* occurring randomly throughout the track.

When the second track is complete, create a new track by selecting the *Tracks→Add New→Stereo Track* menu item. Now select the audio data from the second track and paste it into the newly created track. Use the *Time Shift Tool* to move the new track's audio data horizontally so that it is out of sync with the previous *dinks* track. Repeat this procedure, creating two more tracks into which you paste the second track's audio data. Use the *Time Shift Tool* for each of the new tracks, purposely misaligning the *dinks* in the tracks.

Open the Audacity Mixer Board and reduce the levels of the *dinks* tracks; in addition, pan some of them a bit left and some a bit right. Apply effects to the *dinks* tracks such as GVerb and Delay to add depth to them.

If you performed this exercise correctly, the result should be reminiscent of certain elements of Wendy Carlos' wonderful soundscape "Winter" from her incredible recording *Sonic Seasonings*. If you aren't familiar with this masterpiece, by all means check it out. I swear the room gets colder every time I hear it.

Figure 7-24 shows how my tracks appeared when I completed this illustrative experiment. The top track contains the howling wind sound, and the other tracks contain the high-pitched *dinks*.

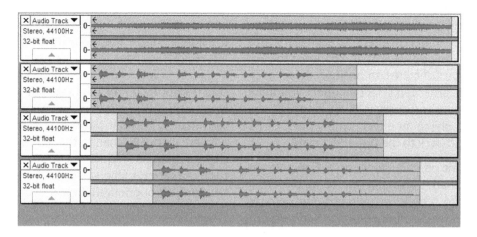

Figure 7-24. *A first attempt at multitrack layering*

Continue to experiment with multitrack layering, adjusting your synth's controls to produce interesting sound layers and then mixing the resulting tracks so that some are dominant and others contribute only subtly to the whole. Here are some scenarios to create using multitrack layering:

- A complex machine
- Factory noises
- A steam engine
- An airplane
- A rocket
- An ethereal soundscape

I hope you're starting to see how these techniques can be combined. For instance, you could create a track using Musique Concrete techniques, which you then include as one of the layers in a multitrack layered sound. The possibilities are endless.

This technique results in very interesting sound layering, but by employing volume envelopes, we can cause the soundscape to change over time. This is the subject of the next section.

Multitrack Volume Envelope Layering

This exercise is similar to the preceding one, and to get started, create another multitrack layered soundscape or use the one you created in the previous exercise. When you have created the new multitrack soundscape, save your work (or, if using the previous one, save the previous project with a new filename). Now let's look at the various methods of applying volume envelopes to our clips.

Figure 7-25 shows the multitrack layered soundscape I created for this exercise prior to application of volume envelopes.

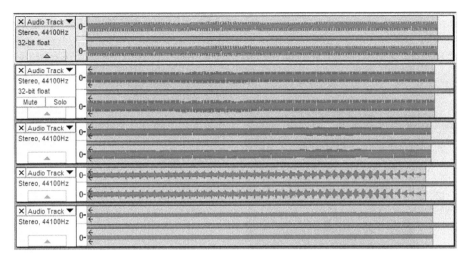

Figure 7-25. Multitrack layered sound (pre–volume envelope application)

Volume envelopes can be applied to clips in two ways. One way uses the *Fade In* and *Fade Out* effects located in the Effects drop-down menu. To use either of these effects, simply select the clip section you want to apply the fade to and select the effect. Figure 7-26 displays how the first track looks after applying *Fade In* to the first third of the clip and applying *Fade Out* to the last third of the clip.

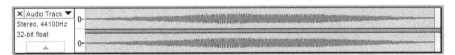

Figure 7-26. Clip after application of Fade In and Fade Out

The second method uses the *Envelope Tool*, which is located on the Audacity Tools toolbar. To use this tool, increase the vertical size of the track to which the volume envelope will be applied in order to have more vertical space in which to adjust its amplitude. When you select the *Envelope Tool*, the tracks change appearance to display the envelope adjustment *handles* (thick blue lines), as shown in Figure 7-27.

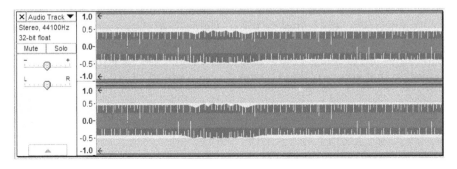

Figure 7-27. Track appearance with Envelope Tool selected

The blue lines at the top and bottom of the clip are the *handles* that you click and drag vertically using the *Envelope Tool* cursor (shown in Figure 7-28) to change the volume at a particular point.

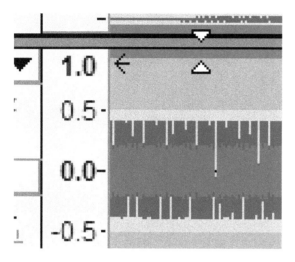

Figure 7-28. *Envelope Tool cursor*

When only one node is created, the whole clip's volume is adjusted if the handle is moved up or down using the *Envelope Tool* cursor. To create an envelope, we must create two or more volume adjustment nodes by clicking at various points along the adjustment handle line with the *Envelope Tool* cursor. Once we've created several nodes, you can move them up or down to create slopes between the nodes, as shown in Figure 7-29.

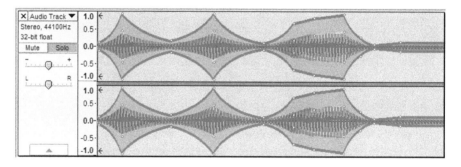

Figure 7-29. *A volume envelope applied to a clip*

Now that we know two ways to apply volume envelopes to our audio clips, let's complete this exercise. Apply both volume envelopes and horizontal shifting to the clips to create a multitrack layered composition that ebbs and flows as the volume of the layered tracks fades in and out. Figure 7-30 shows what my project looked like after applying volume envelopes using both the Fade In–Fade Out method and the *Envelope Tool* method.

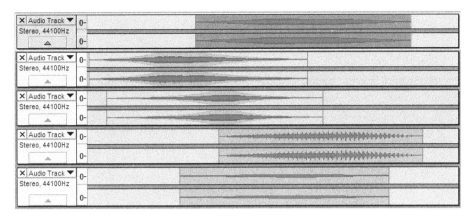

Figure 7-30. *Multitrack volume envelope layering composition*

To complete this exercise, apply volume envelopes to the multitrack soundscapes you produced in the previous section or create an entirely new one. A favorite soundscape to produce in this manner is *The Factory*, in which a variety of industrial sounds come and go via volume envelopes and panning. A variation on this theme is *The Spaceship*, in which the composer imagines he is walking through a large spaceship where a wide variety of sounds occur as gears whir, instruments beep, and gigantic engines throb.

Floobydust

The sound generation possibilities available to you by combining the techniques discussed in these creativity exercises are limitless. I also encourage you to investigate the use of *SFZ* files. The *SFZ* file format, in combination with an SFZ-aware soft-synthesizer (often included with DAWs), allows you to use short WAV or OGG sample snippets and play the sound with your MIDI keyboard. If you begin to experiment with *SFZ* files, you might find the MFOS Web SFZ Helper application (*http://bit.ly/17UDoop*) useful. It provides an interface with familiar slide controllers with which you set the various *SFZ* file parameters. It translates the interface's control settings to valid *SFZ* file content.

There are many websites and discussion forums dedicated to synth-DIY. Using search terms like *DIY synth*, *analog synthesizer*, or *electronic music* will definitely return a long list of websites to visit. A few sites I'm familiar with are:

- electro-music.com
- Synthtopia (*http://www.synthtopia.com/*)
- Analogue Haven (*http://www.analoguehaven.com/*)
- The email forum for Analogue Heaven (*http://machines.hyperreal.org/Analogue-Heaven/*)

I have thoroughly enjoyed sharing all of this great analog synthesizer and synth-DIY information with you, and I sincerely hope that it puts you firmly onto the road of having creative fun making your own analog noise boxes and music synthesizers. If there were a degree program for synth-DIY, you would be ready to graduate with honors. I sincerely hope that all of your adventures in synth-DIY are positive, educational, and very, very successful.

So what's floobydust? You'll just have to get a copy of the *1980 National Semiconductor AUDIO/RADIO Handbook* and read Chapter 5 to find out. Namasté y'all.

A Field Guide to Op Amp Circuit Applications

In this appendix, I'll present a number of useful analog synth circuit building blocks, including inverting and noninverting buffers, comparators, integrators, and digital-to-analog converters. For each circuit element, I'll present the schematic and explain how the circuit works. Many of these circuits have generally useful functions that can be put to work in your synth-DIY work, as well as your electronic projects.

Buffers

Here are some circuits you can use when you need to boost the level of a signal. Typical op amp outputs can only source or sink about 4 to 5 mA before the output begins to degrade and distort. When driving LEDs with an op amp, this current drive limit isn't a big deal, but when it comes to the signal chain, avoid overtaxing op amps. Sometimes a design will only require three op amps from a quad JFET or other high-impedance input op amp package. Be sure to keep any unused op amp quiet by connecting its output to its inverting input and its noninverting input to ground. Otherwise the inputs could be influenced by leakage currents or electrical fields, resulting in oscillation at the unused op amp's output and excessive current consumption.

Inverting Buffer

Figure A-1 is the circuit from Figure 5-8 presented with a table of resistance values you can use to set gains. That's a .1µF AC coupling cap on the input and a 10µF aluminum bipolar (also called nonpolarized) AC coupling cap on the output. You can set any gain you want by observing the gain formula, which is the quantity (RFB value divided by RIN value) multiplied by –1.

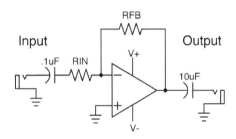

Figure A-1. Generic inverting buffer

Table A-1. Setting the gain of the inverting buffer

RIN Value	RFB Value	Resultant Gain
10K	100K	–10
20K	100K	–5
47K	100K	–2.13
100K	100K	–1
100K	10K	–.1

Noninverting Buffer

Figure A-2, a copy of Figure 5-6, is an op amp configured as a noninverting buffer. That's a .1µF AC coupling capacitor on the input and a 10µF aluminum bipolar (also called nonpolarized) AC coupling cap on the output. You can set any gain you want by observing the gain formula, the quantity (RFB value divided by RG value) + 1. Table A-2 presents some example resistor values along with the resulting noninverting gains. Hey, what's that 1M resistor doing there? Op amps aren't perfect, and any offset current from the noninverting input will charge the coupling cap, causing an unwanted change in the DC level of the op amp's output; this can cause positive or negative saturation. The 1M resistor pulls down that side of the capacitor to ground. When choosing values for the noninverting input's coupling cap and resistor to ground, remember that if the cap is too small or the resistor to low in value, you will be creating a high-pass filter, which will alter the signal you apply to the noninverting input.

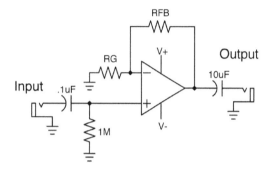

Figure A-2. *Generic noninverting buffer*

Table A-2. *Setting the gain of the noninverting buffer*

RG Value	RFB Value	Resulting Gain
10K	100K	11
20K	100K	6
47K	100K	3.13
100K	100K	2
100K	10K	1.1

High-Impedance Buffer Follower

The circuit in Figure A-3 shows an op amp in a special case of the noninverting configuration. The output is directly connected to the inverting input, and the input voltage is fed directly to the noninverting input. Here's why: sometimes you need to measure a voltage in a situation where you don't want to affect it by loading it or applying any type of filtering to it. In short, you don't want to change the voltage you're trying to measure any more than you have to. The gain of the op amp in this configuration is always 1, since any voltage applied to the noninverting input is immediately matched by the output of the op

amp to keep the inverting input at the same voltage.

High Impedance Buffer Follower

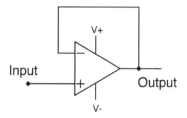

Figure A-3. *High-impedance buffer follower*

The impedance of the noninverting input of a JFET input–type op amp can be tens to hundreds of megaohms, which is kind of like touching whatever you want to measure with a little breath of air. The input capacitance of a JFET op amp is typically a few picofarads, so even when looking at a high-source impedance, the noninverting input is polite and unobtrusive. The upshot of the whole thing is that you can connect the noninverting input to whatever voltage in your circuit you want to look at, and that part of the circuit will hardly notice. For example, a high-impedance buffer follower circuit can be used in a sample and hold circuit to buffer the voltage stored on a relatively small cap for tens of seconds before the voltage starts to leak either through the cap or through the input offset current of the op amp. Keep this circuit in your bag of tricks.

What Is Biasing?

Don't be intimidated by the term bias or biasing; it simply means to add a positive or negative voltage offset (or bias) to a circuit node. Most synthesizer modules use operational amplifier ICs requiring a bipolar power supply. This is a power supply that outputs positive and negative voltage in addition to ground (or neutral). This type of supply facilitates circuit design in which op amp *outputs oscillate **about ground**; the op amp's output amplitude ranges above and below zero volts. However, if we need to limit the op amp's output excursions to the positive voltage realm, we must **bias the op amp's output up** so that it oscillates about some more positive voltage. The opposite can be true as well, and in that case, we would bias the op amp's output down.*

Comparators

Following are some schematics for op amp comparator circuits that you will find useful in your synth-DIY work.

A Basic Comparator

Often you need to light an LED when a voltage exceeds a certain level. The comparator circuit shown in Figure A-5 is just the ticket. When the op amp's output is at positive saturation, the LED will be forward biased and light up. You can change the brightness by adjusting the value of R3 lower for more brightness and higher for less. The VRef threshold voltage can be adjusted by setting the values of the resistors used for the biasing voltage divider.

Biasing the Op Amp's Output

There will be times when you want to *bias* or add a positive or negative DC offset to an op amp's output. Figure A-4 shows one way to accomplish this goal.

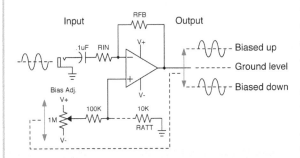

Figure A-4. *Op amp output biasing*

Notice that we remove the AC output coupling cap shown in previous illustrations because if we didn't, the capacitor would block the DC bias we added to the op amp's output. An example of when this would make sense is if we were applying the op amp's output to the input of a CMOS CD4066 analog switch. The CMOS CD4066 analog switch operates from a maximum supply voltage of ground and +15V, whereas the op amp's normal output would be oscillating about (equally above and below) ground. Putting negative voltage into the input of a CMOS gate will cause its internal protection diodes to conduct and bring about all manner of mysterious and confusing circuit operation. In order to raise the DC output voltage level of the op amp to ensure that it only outputs voltage in the positive half of the world, we would bias the output up.

The inverting gain formula still applies to the op amp in this configuration. Only now, instead of the noninverting input being connected solidly to ground, we are raising or lowering the voltage on it, which in turn affects the output voltage for all of the reasons we discussed above in the section about op amp operation. Bear in mind that the voltage applied to the noninverting input will be multiplied by the gain set by RIN and RFB. Therefore there will be times when you need to considerably attenuate the voltage applied to the noninverting input. That's the purpose of 10K RATT (R Attenuate), which will divide the voltage applied to the noninverting input between the 100K resistor and itself. Thus, for a high gain situation you will want to add the attenuating resistor and adjust its value depending on how much gain the op amp is providing. Higher gain scenarios will require a lower value RATT, and lower gain scenarios will require a higher value for or the removal of the RATT.

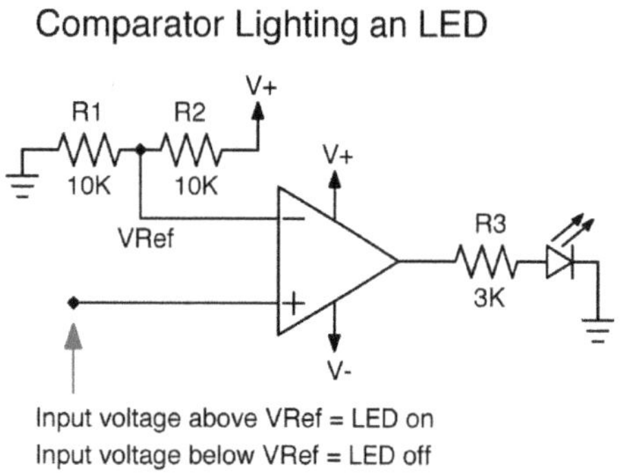

Comparator Lighting an LED

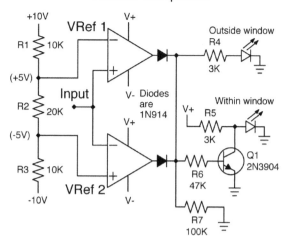

Input voltage above VRef = LED on
Input voltage below VRef = LED off

Figure A-5. *Comparator lighting an LED*

A bit of Ohm's law will lead you to the values you need for a particular voltage level. Don't go too low in the values you select for the voltage divider or it will needlessly drain excessive current. If you need the threshold level to be below ground, you can connect resistor R2 to V- instead of V+ and then just compute the values you need to achieve the voltage you want. You can swap the inverting and noninverting input connections to achieve an inverting version of this circuit. By doing that, the LED will turn on when the comparator's input voltage (which is applied to the inverting input this time) goes below the threshold voltage level.

The Window Comparator

Occasionally you need a comparator to let you know when a voltage is within (or outside of) a range of voltage, and the window comparator is just the tool for the job. The schematic is shown in Figure A-6, and you can see that the window comparator is actually two comparators in one. The *window* is the range of voltage between VRef 1 and VRef 2. VRef 1 is applied to the inverting input of the top comparator and VRef 2 is applied to the noninverting input of the bottom comparator.

Window Comparator

Figure A-6. *Window comparator with LEDs*

The input is connected to both the top comparator's noninverting input and the bottom comparator's inverting input. When the input's voltage is above VRef 1, the output of the top comparator goes to its positive saturation voltage. When the input's voltage is below VRef 2, the output of the bottom comparator goes to its positive saturation voltage. When the input voltage is within the range of VRef 1 and VRef 2, the output of both comparators go to their negative saturation voltage.

The outputs of both comparators are connected to the anodes of two diodes whose cathodes are connected. If either comparator goes high (positive saturation voltage), the *Outside window* LED illuminates. When both comparator outputs are low (negative saturation voltage), both diodes are reverse-biased, and the *Outside window* LED goes off. The *Within window* LED illuminates when both comparator outputs are low (input voltage within window) because there is no current to turn on Q1, and its base is held at ground by 100K resistor R7. Consequently, the *Within window* LED lights because current can flow through the R5 3K resistor and illuminate the LED. When Q1 is on as a result of either comparator's output being high (*Input outside window*),

Q1 shunts the current from R5 to ground and turns off the *Within window* LED.

The window's voltage range does not have to be symmetrical, as shown here, but can be biased in one direction or another and can be widened or narrowed as needed. If R1 and R3 were both 20K and R2 was 10K, the window would be narrower. If R2 was made higher in value and R1 and R3 remained the same, the window would become wider.

Signal Mixing

Just about all analog synth modules accept various types of inputs, be they signal inputs or modulating control voltages. The way the input signal is coupled to the module's circuitry is very important, with some inputs being DC coupled and some being AC coupled.

For example, a VCO's exponential control voltage input is DC coupled. When we apply a sustained level of DC voltage to a VCO, we expect the oscillator to stay at the frequency associated with that voltage until we change it, no matter how much time goes by between control voltage changes. If the VCO's exponential control voltage input was AC coupled, meaning applied through a capacitor instead of directly or through a resistor, only the rising or falling edges of the voltage changes applied to the input would get through. This would produce a momentary rising or falling chirp from the VCO in response to CV changes, but after that, the VCO would return to the frequency it had been set to by its coarse and fine frequency adjust knobs.

On the other hand, we want AC coupling for the inputs to mixers, filters, and most signal processors because it removes any DC offset that might be present on the signal to be processed. An input signal with DC offset can significantly change the operation of a signal processor if it is not blocked by AC coupling. This is because any gain applied by the processor would also amplify the DC offset and could result in unwanted clipping at the output of the signal processor.

Simple AC-Coupled Audio Mixer

Figure A-7 shows a simple schematic you can use for all your mono-mixer needs. The input signals are each applied via an input jack whose tip terminal is connected to one end of a 10K volume (or level) potentiometer. The potentiometer's other end goes to ground, and its wiper connection connects to an AC coupling cap (.1µF in illustration). Each AC coupling cap goes to a corresponding 100K input resistor, which finally connects to the op amp's inverting input. Notice that the feedback resistor is 100K, and since the input resistors are all 100K, the op amp will apply a gain of 1 to each of the AC input signals.

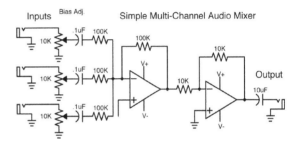

Figure A-7. *Simple audio mixer*

The output of the mixing stage is DC coupled to an inverting buffer with a gain of 1, which reinverts the output of the first inverting stage. The output of the second inverting buffer connects to the output jack via a 10µF aluminum bipolar (also called non-polarized) AC coupling cap. I challenge you to hear the difference between a signal and its inverted version, but keeping all of the signals in their original form is usually a good practice. When a signal and its inverse are added, depending on their individual levels, they can cancel one another partially or completely. You can add as many inputs (jack, pot, coupling cap, and input resistor) to the circuit as you like.

Another advantage of the first inverting buffer mixer is that, in this configuration, the inverting input is referred to as a "virtual ground" because, as the noninverting input is connected to ground and we know the output of the op amp is constantly trying to balance the voltage on the two inputs, the signals on the input do not interact with one another as they would in a passive mixing scheme (no op amp used). It is as if each of the signals applied to the input were simply and separately connected to a 100K load resistor to ground (*virtual* ground, that is).

It is easy to change the overall gain of the mixer by changing the value of the feedback resistor in the second stage inverting buffer. Use the gain formula shown in Figure 5-5 to determine the values for the desired gain. You can also change any individual channel gain in the first mixing buffer stage by adjusting the value of any of the 100K input resistors (lower value, higher gain; higher value, lower gain). Again, Figure 5-5 applies.

Simple Multichannel DC Modulation Mixer

Figure A-8 looks very much like Figure A-7, but we have removed the input and output coupling capacitors. Now we can use this mixer for mixing the outputs of modules whose voltages change far too slowly to be effectively coupled through a capacitor. Modulators such as LFOs, sample and holds, ADSREGs, etc., are designed to be DC coupled to the module they are modulating.

Simple Multi-Channel DC Modulation Mixer

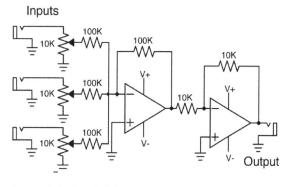

Figure A-8. *Simple DC modulation mixer*

This mixer works just as you would expect and mixes and inverts the signals in the first op amp stage, then reinverts them via the second. The

mixer provides a gain of 1 for the DC signals connected to the input. The potentiometers used to adjust the input levels help you mix signals of various amplitudes. However, you could easily drive the op amps into saturation if high-level signals are applied and the input levels are turned up too far. A cool thing to do would be to add the window comparator presented above to the output and have a red LED illuminate if the signal falls outside of a preset voltage window. You can add as many inputs (jack, pot, and input resistor) to the circuit as you like.

Simple but Useful Op Amp Oscillators

You can never have enough ways to make an oscillator, and there are hundreds of ways to go about it. I'm going to show you a simple, single op amp square-wave oscillator and a second oscillator that requires two op amps, but pays us back for the trouble by delivering two waveforms, square and triangular. Along the way, we'll go over using the op amp as a ramp generating integrator, yet another cool op amp application.

Single Op Amp Square-Wave Oscillator

Looking over Figure A-9, we see that we have two feedback paths connecting the output to *both* inputs. What's *that* all about? I'll explain. We will consider that V+ and V- in the circuit are positive and negative 12V, respectively. The way we have to approach understanding this circuit is to know that when power is applied, the op amp's output will pretty much arbitrarily shoot to either positive or negative saturation because the positive feedback through R4 guarantees it. So flipping a coin, we'll say that the output starts up in high saturation.

Single Op Amp Square Wave Oscillator

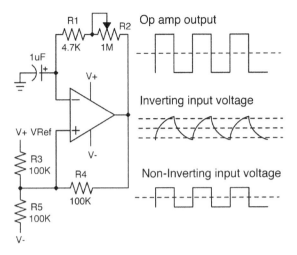

Figure A-9. *Single op amp square-wave oscillator*

The noninverting input is connected to three resistors, R3 (100K to V+), R5 (100K to V–), and R4 (100K to the output). If it weren't for the current coming through R4, the noninverting input would be held at 0V by the voltage divider formed by 100K resistors R3 and R5. But current does come through R4, and when the output of the op amp is high, the voltage on the noninverting input will rise to roughly 4V. The 1µF cap hanging off the inverting input is discharged prior to power application, so when power is applied and we come up in high saturation (as explained above), the inverting input is still at ground (discharged) level, but the noninverting input immediately attains roughly 4V.

Since the 1µF cap has to charge from ground, the output remains high until the cap charges to a voltage that is slightly above the roughly 4V sitting on the noninverting input. At that time, since the inverting input is seeing a higher voltage than the noninverting input, the output of the op amp slams to low saturation. Two things happen then: one is that the noninverting input immediately jumps to roughly -4V, and the second is that the current fed back through R1 and R2 begins to discharge the 1µF cap. When the 1µF cap discharges to a level slightly below the volt-

age on the noninverting input (roughly -4V), the voltage on the inverting input will be lower than the voltage on the noninverting input, and the op amp output will slam to positive saturation.

This process continues producing square-wave oscillation on the output of the op amp. By adjusting the 1M pot in the inverting input's feedback path, you can speed up (lower resistance in the capacitor's charge path) or slow down (higher resistance in in the capacitor's charge path) the oscillators frequency. You can also change the value of the 1µF capacitor to affect the range of the oscillator's adjustment. Higher values of capacitance will result in lower frequency range and lower values just the opposite. Pretty snazzy, huh?

Figure A-9 also shows 1) the op amp's output at the top oscillating between positive and negative saturation voltage, and 2) the voltage on the inverting input as the cap charges and discharges to the threshold level set on 3) the noninverting input, which is shown in the bottom waveform.

Two Op Amp Square- and Triangle-Wave Oscillator

Figure A-10 shows op amp U1 in a configuration we have not yet discussed. When you see an op amp with nothing in its inverting feedback path but a capacitor (like in U1's case), the op amp is being used as an *integrator*. When in this role, the op amp *integrates* the current flowing into or out of its inverting input and reflects the result by outputting an *area-under-the-curve*-shaped voltage envelope. In our previous discussions about negative feedback, we always had a resistor in the feedback path and could easily see how current from the op amp's output could influence the voltage on the inverting input. When we have a capacitor in the feedback path, the current still flows, but it flows *through* the feedback capacitor.

Two Op Amp Square And Triangle Wave Oscillator

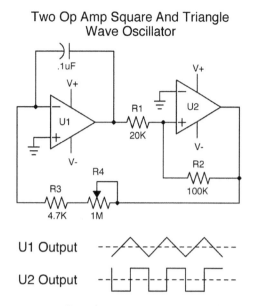

Figure A-10. *Two op amp square- and triangle-wave oscillator*

Wait a minute, I thought capacitors blocked DC. That's true, they do—so the only way to get current to flow through a capacitor is to maintain a voltage differential across its plates. The op amp's output comes to the rescue again! To get current to flow through the feedback capacitor, the op amp's output voltage, in response to a constant current flowing into or out of its inverting input, falls or rises linearly at a rate that causes the same current to flow through the feedback capacitor as is flowing into or out of the inverting input. The operation of op amp U1's output is the same, regardless of the fact that we are using a capacitor for its feedback element instead of a resistor. U1's output adjusts to push or pull (via ramping up or down) enough current through its .1µF feedback cap to maintain the same voltage on its inverting input as is applied to its noninverting input (ground in this case), regardless of changes in voltage on the inverting input.

If current continues to be pulled away from the inverting input, the output would ramp up until the op amp's output was in positive saturation. The opposite would also be true: thus when an integrator is used, you need a way to *reset* the output to zero or control its voltage excursions

so that they don't go to positive or negative saturation voltage. The Noise Toaster's VCO uses an N-Channel JFET to short the integrator's feedback capacitor as a means of returning the output to ground (see "The Voltage-Controlled Oscillator (VCO)" on page 99 in Chapter 6). The circuit presented in Figure A-10 controls the ramp's excursions.

The magnitude of the current flowing into or out of the inverting input in this configuration determines the rate at which the voltage on the op amp's output ramps down or up, respectively. Low current results in a slower ramp, and higher current results in a faster one. The size of the capacitor is also a huge factor in the rate of the integrator's ramps. Large capacitors have more plate area and thus can pass more current with a slower ramp, whereas small capacitors with less plate area need a faster ramp to conduct the same current.

Whew! Sorry about that detour, but I really think you'll understand the explanation of Figure A-10 a lot better now, so here we go. In this schematic, we have an op amp used as an integrator (U1) and one used as a comparator (U2). When we apply power to this circuit, the output of U2 will come up either saturated high or saturated low because we have no negative feedback, and the positive feedback provided by R2 ensures it.

For this explanation, we'll imagine that U2's output came up saturated low. On power-up, the .1µF feedback capacitor between U1's inverting input and its output will be discharged, and the output of U1 will be at or very close to ground. Since U2 came up saturated low, we will be pulling current away from the inverting input of U1 through R3 (4.7K resistor) and R4 (1M potentiometer) in series. In response, the output of U1 will start to ramp up linearly in the positive direction as it pushes a steady stream of current through the .1µF cap into U1's inverting input to balance the current being pulled away via R3 and R4. The output voltage of U1 will continue to ramp up until it pushes enough current through R1 (20K resistor) to surpass the current being drawn away by U2's output via positive feedback resistor R2 (100K resistor).

As soon as the voltage on U2's noninverting input goes slightly above the voltage on its inverting input (ground in this case), the output of U1 shoots to high saturation. Now that the output of U2 has gone high, current will be pushed toward the inverting input of U1 via R3 and R4, and the output of U1 will reverse direction and begin to ramp down, pulling current through the .1µF feedback capacitor in order to keep the voltage on U1's inputs balanced. When U1's output voltage ramps low enough to pull sufficient current through R1 to surpass the current being sourced from U2's output via R2, U2 shoots to low saturation, and the cycle repeats. See, it's not so hard to understand when you know what the op amps are doing.

The 1M potentiometer (R4) in the path between the output of U2 and the inverting input of U1 adjusts the rate of the square/triangle oscillator. When you adjust it for less resistance, more current is able to reach U1's inverting input, and the output of U1 has to rise or fall faster to push enough current through its feedback capacitor to balance it. When you adjust R4 for more resistance, less current reaches U1's inverting input, and the output of U1 rises more slowly since it is compensating for less current.

What will the amplitude of the ramp wave be? If we consider that we are powering the op amps with plus and minus 12V the magnitude of the saturated op amp output will be approximately 10.5V positive or negative. The current flowing toward U2's noninverting input when U2's output is saturated high will be about 105 µA. In order to get 105 µA to flow away from the noninverting input through R1 (20K), we need –2.1V. Thus the output of the triangle-wave generator will ramp down to slightly below –2.1V when U2 is saturated high and ramp up to slightly above 2.1V when U2 is saturated low, providing a triangle wave that is slightly greater than 4.2V peak to peak in normal operation.

How would we make the triangle wave have more amplitude if we wanted to? We would add more hysteresis by either increasing the value of R1 or decreasing the value of R2. Conversely, to reduce the triangle wave's amplitude, we would reduce the hysteresis by either decreasing the value of R1 or increasing the value of R2.

Digital-to-Analog Conversion Using Binarily Weighted Resistors or the R/2R Resistor Ladder

When you want to convert binary data to analog voltage levels, you need to use a DAC or digital-to-analog converter. You can make your own DACs using resistors and op amps for your synth-DIY work. The Music From Outer Space Voltage Quantizer uses DACs made from resistors and op amps to convert continuous analog voltage levels to digital data and then back to stepped, discrete analog voltage levels. You can feed in whatever voltage you want, and the Voltage Quantizer converts it to the nearest discrete voltage level. The discrete voltage level's output by the Voltage Quantizer can be set to be half-note increments (83.33 mV steps), whole-note increments (166.6 mV steps), or four half-step increments (333.32 mV steps). With DACs, you can make cool voltage step generators to control any of your voltage-controllable modules.

Please Feed the DAC Inputs from CMOS Outputs

The DACs shown here expect to be fed from CMOS data lines. CMOS sources and sinks current, has V+ and ground level outputs, and provides consistency between outputs. See **Appendix C** for more on CMOS components.

The Binarily Weighted Resistor DAC

The binarily weighted resistor DAC is shown in Figure A-11. It works by applying the digital data lines to resistors, where each resistor is half the value of the previous one in the ladder. Notice that data line A's (LSB) resistor value is 80K; data line B's resistor value is half of that, or 40K; data line C's resistor value is half of that, or 20K; and lastly, data line D's (MSB) resistor value is half of its predecessor, or 10K. When any one of the data lines is high (+12V), current flows through the weighted resistor into the inverting input of U1-A, resulting in a negative voltage at its output. When a data line is low, it does not contribute current to the input of U1-A, which acts as an inverting summer.

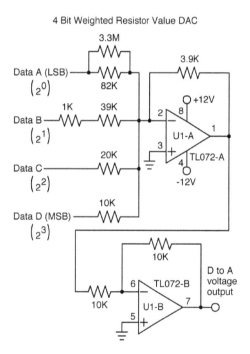

4 Bit Weighted Resistor Value DAC

Figure A-11. *DAC using weighted resistor values*

The feedback resistor of U1-A determines the size of the voltage steps output by the DAC. You need to ensure that the feedback resistor value is not too high, else the current from the inputs at high counts will cause U1-A to enter negative saturation. With the values and supply voltage shown,

if only data line A is high, the voltage at the output of U1-A will be –0.585V; if only data line B is high, –1.17V (notice this is twice as high); if only data line C is high, –2.34V (notice this is again twice as high); and finally, if only data line D is high, –4.68V (again, twice as high). By applying any binary count between 0 and 15 to the inputs, the voltage output by U1-A will be the binary count times the LSB voltage of 0.585V times –1. This voltage is inverted by U1-B acting as a gain of one inverter so that increasing count equals increasing voltage at U1-B's output.

You can experiment with this circuit on a solderless breadboard by feeding in the Q1 through Q4 outputs of a CD4024 binary counter, clocked by the simple CD40106 oscillator circuit presented in Appendix C. Don't forget to connect the CD4024's reset line to ground.

You can use more data lines if you need to. Just remember that the least significant data line's resistor always has the highest value, and each successively higher data line's resistor value is equal to half of its predecessor's value. So for example, if you had eight data lines and used a 1M resistor on the LSB data line, you would need resistors with the following values for the successively higher data lines: 500K, 250K, 125K, 62.5K, 31.25K, 15.625K, and 7.8125K. If you think you might have a problem finding these resistor values, you're probably correct. Herein lies one of the problems of the weighted resistor DAC. Finding or cobbling together binarily weighted resistor values can be a real pain. The R/2R resistor ladder presented next is far easier to find resistor values for, as you'll soon see.

The R/2R Resistor Ladder DAC

The R/2R resistor ladder concept is a piece of true genius (see Figure A-12). Try as I might, I was unable to find a definitive inventor for the concept, though it has been in use for a long, long time. I'm sure that if you spent an afternoon and worked it all out using Kirchoff's and Ohm's laws, as well as a healthy dose of superposition, you'd understand exactly how it works. All I know is that it works perfectly and that you only need two resistor values to do it: R and 2R.

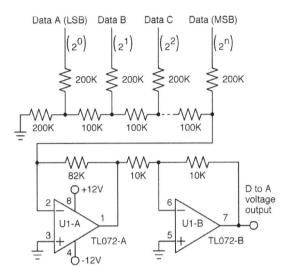

Figure A-12. *DAC using R/2R resistor ladder*

As you can see from the schematic, the data lines are fed to the 2R value resistors. The other ends of the 2R value resistors have the R value resistors strung between them. The opposite end of the 2R value resistor used in the LSB position also has a 2R resistor connected to it, which goes to ground. The DAC's output is taken from the junction of the MSB 2R value resistor and its connected R value resistor. This is the pattern, and you can string it out for as many bits as you want and you'll still only need two values of resistor. I chose 100K and 200K for my DAC because they're relatively high impedance, which keeps the CMOS counter outputs feeding them happy. By not loading the outputs of the CMOS counter significantly, each one delivers the same high and low logic voltage levels. Remember that if you load the output of a CMOS device too heavily, you will affect the voltage of its logic levels. Pulling too much current from a high CMOS output will tend to lower its logic 1 voltage. Sinking too much current into a low CMOS output will tend to raise its Logic 0 voltage. Using relatively high resistors in your R/2R DAC will minimize or eliminate these

effects and ensure consistent and accurate digital-to-analog conversions.

I chose to apply the output of the R/2R resistor ladder DAC directly to the inverting input of an op amp. Therefore the current flowing out of the DAC network is converted to a voltage at the output of U1-A. As the binary count increases, the voltage goes down in equally sized voltage steps. We invert the output of U1-A using U1-B as an inverter with a gain of one. As in the case of the binarily weighted resistor value DAC, the feedback resistor of U1-A has to be chosen so that the full count does not result in negative saturation at its output. The selection of 82K keeps the output well with the necessary range. By adjusting the value of U1-A's feedback resistor, you can tailor the step size for your application's need.

Regardless of the type of DAC used, the precision of the resistors is very important. For the weighted resistor value DAC, the 2-to-1 relationship between successively higher bits is the important thing. For the R/2R ladder DAC, you just need to make sure the you have the 2-to-1 ratio between the resistor values used. Use metal film resistors of 1% or better tolerance. In the R/2R DAC circuits in my own gear I use 0.1% tolerance 100K and 200K resistors (available on the Music From Outer Space website).

Light Up Your Synth with LEDs

Every analog synthesizer looks way cooler when it has LEDs flashing all over the place, so in this section, I'm going to show you several ways to make that happen. LEDs provide a nice visual indication of power on status, gate status, voltage levels, LFO rate, sample and hold rate, active sequencer channel, etc. LEDs are semiconductor diodes whose dopants give them the amazing ability to emit visible light under forward bias. Originally only available in red, today LEDs come in a variety of wavelengths, ranging from invisible infrared to a wide assortment of vibrant, visible colors. The examples show +12V/−12V, but your circuit may use a different voltage, which is perfectly fine. At lower voltages, use a lower value for the LED's current limiting resistor; and at higher voltages, you may want to use a bit higher value. What do you say we light something up?

The Simplest Way to Light an LED

Figure A-13 presents the simplest way to light an LED. Connect the LED's current limiting resistor to V+, and connect the other end of the resistor to the LED's anode and the LED's cathode to ground. When power is applied, current flows through the current-limiting resistor and forward-biased LED, causing it to glow. The lower the resistor value, the brighter the LED. But if you go too far, you can destroy the LED, so keep the current-limiting resistor in the 300 to 3K range, depending on power supply voltage and your taste in LED brightness. When using a 5V supply, 330 ohms is a popular resistor value to get your LEDs to blaze. For 9V to 15V, I like to stay in the 1K to 3K range, depending on the efficiency of the LED used and how bright I want it to glow.

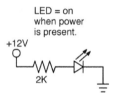

Figure A-13. *LED power indicator*

Isolating the LED Control Source with a Transistor (I)

In Figure A-14, we are isolating the relatively high current needed by the LED from the controlling source by using a transistor. To turn the LED on, we apply +12V via the 47K base drive resistor, which causes current to flow from the supply through the 2K current limiting resistor, through the transistor, and finally through the LED, turning it on. In this configuration, when the base drive resistor is connected to ground potential, the LED is off; and when it is connected to +12V, the LED is turned on.

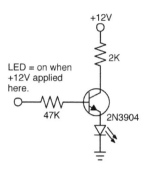

Figure A-14. *Light an LED using a transistor*

Isolating the LED Control Source with a Transistor (II)

The circuit shown in Figure A-15 also isolates the LED drive current from the controlling source with a transistor. However, in this circuit, the LED is lit when ground potential is applied to the base of Q1 via its 47K base drive resistor. With the base drive resistor for Q1 grounded, Q1 is off, allowing current to flow to the base of Q2 via the 47K resistor to V+, which turns Q2 on. When Q2 is on, current flows from +12V through the 2K current-limiting resistor through Q2 and lights the LED. If greater than about a volt is applied to Q1's base via its 47K base drive resistor, it turns on. When Q1 turns on, the current that was formerly driving the base of Q2 via the 47K resistor to +12V is shunted to ground through Q1, turning Q2 off. When Q2 is off, no current flows through the LED via the current-limiting 2K resistor, and thus it is off.

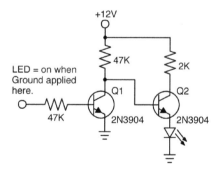

Figure A-15. *Light an LED using two transistors*

Isolating the LED Control Source with a CD40106 Inverter (I)

Figure A-16 shows how an LED can be driven simply using a CD40106 Hex Schmitt Inverter, which we cover in Appendix C. Benefits of doing it this way include very high input impedance, the inverter's inherent hysteresis, and convenience. In this circuit, the LED lights when the input to the inverter is at logic 1. Since a logic 1 on the inverter's input results in a logic 0 on its output, current flows from +12V through the LED and 2K current-limiting resistor into the inverter's current sinking input, lighting the LED. When the input to the inverter is low, the output is high, and the LED is reverse-biased and thus off.

Logic 0 = LED off
Logic 1 = LED on   +12V

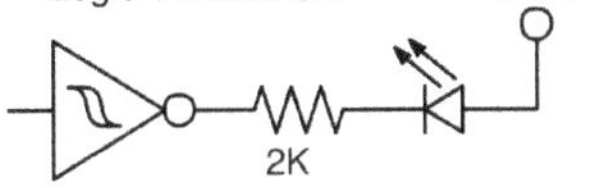

Figure A-16. Light an LED using CD40106 (I)

Isolating the LED Control Source with a CD40106 Inverter (II)

The circuit shown in Figure A-17 is similar to the one just presented, except that when the input to the inverter is logic 1, the LED is off, and vice versa. The LED lights when the input to the inverter is at logic 0. Since a logic 0 on the inverter's input results in a logic 1 on its output, current flows from the inverter's output through the 2K current-limiting resistor through the LED to ground, lighting the LED. When the input to the inverter is high, the output is low, and the LED is reverse-biased and thus off.

Logic 0 = LED on
Logic 1 = LED off

2K

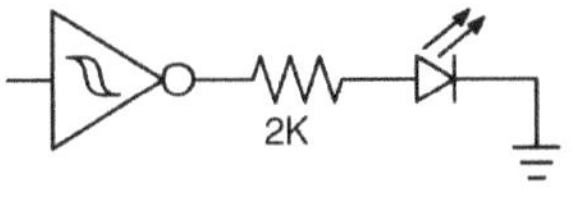

Figure A-17. Light an LED using CD40106 (II)

Light an LED Using a Comparator (I)

When we need to light an LED in response to a voltage exceeding a certain threshold, doing it with a comparator is the natural choice. In Figure A-18, we have an op amp configured as a comparator (*no negative feedback resistor is a giant giveaway*). The comparator's threshold voltage is set by the resistor divider comprised of two 10K resistors. Since one side connects to +12V and the other to ground, 6V is being applied to the comparator's inverting input. If the voltage on the noninverting input is below 6V, the output of the comparator will be saturated low, and the LED will be reverse-biased and thus off. However, if the voltage applied to the noninverting input goes above 6V, the comparator will go into positive saturation, and current will flow through the 2K current-limiting resistor, forward biasing and turning on the LED.

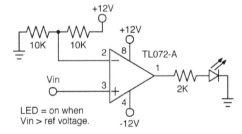

Figure A-18. Light an LED using a comparator (I)

You can, of course, make the comparator's threshold voltage whatever you need it to be using the proper resistor values in the threshold voltage-producing divider. To do so, calculate the current through both divider resistors (+V/(divider resistor 1 + divider resistor 2)), and then multiply the result by the value of the resistor that connects to ground to get the threshold voltage. For example, +12V/(10K + 10K) = .0006A. Then take .0006A x 10K to get 6V.

As a breadboard exercise, what happens if you turn the LED around so that the anode is at ground and the cathode faces the comparator's output?

Light an LED Using a Comparator (II)

This circuit in Figure A-19 is a repeat of the previous one, but notice that we've switched the role of the inverting and noninverting inputs. Now the threshold voltage is applied to the noninverting input. Whenever the voltage on the inverting input is *below* the threshold voltage, the LED will light because the comparator's output will go into positive saturation when the voltage presented to the inverting input is below that presented to the noninverting input. Again, you can set the comparator's threshold voltage to whatever you need by using the appropriate re-sistors in the divider. What happens when you turn the LED in this circuit around, anode to ground? Where's that solderless breadboard?

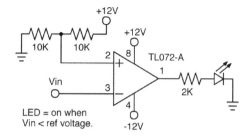

Figure A-19. *Light an LED using a comparator (II)*

The LM13700

The LM13700 Dual Transconductance Op Amp is one of the coolest integrated circuits around today for building the components of analog synthesizers. A *transconductance op amp* is a special type of op amp that can be thought of as a voltage-controlled current source. The LM13700 is one of the few really useful transconductance amplifiers that are still in production.

Several chip houses are manufacturing the LM13700 with different part numbers, but all use the same circuit originally designed by Don Sauer and Bill Gross, of National Semiconductor. Read Don Sauer's interesting account of the chip's design and development on his website, **The LM13600/LM13700 Story** (*http://www.idea2ic.com/LM13700.html*).

The LM13700 is at the heart of many VCFs, VCAs, VCO wave shapers, panners, electronic volume controls, and sine wave oscillators. In "Three Examples of LM13700 Circuits" on page 143, I present some simple LM13700-based electronic music circuits: a VCA, VCF, and a triangle-to-sine wave converter that you can experiment with on your solderless breadboard.

The LM13700 contains two current-controlled transconductance amplifiers that are nicely matched, since they are on the same substrate and thus at the same temperature. This is an important quality when using the LM13700 in the types of resonant filters we're interested in, since they need to function as a matched pair. Additionally, there are two high-impedance Darlington buffer amplifiers available on the chip. The transconductance op-amp's schematic symbol highlights its op amp differential inputs, linearizing diodes, controllable current source/sink output, and Darlington buffers. Figure B-1 shows the pinout of the LM13700 integrated circuit. Table B-2 details each pin's function.

LM13700 Dual Transconductance Operational Amplifier

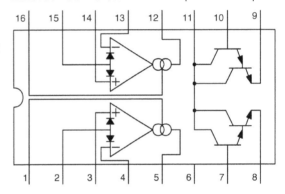

Figure B-1. *LM13700 Dual Transconductance Op Amp pinout*

LM13700 Manufacturers

Table B-1 contains a list of companies producing the chip and the part number(s) used by each.

Table B-1. LM13700 manufacturers

Texas Instruments (previously National Semiconductor)	LM13700 and LM13600
ON Semiconductor	NE5517
New Japan Radio (JRC)	NJM13600 and NJM13700
CoolAudio	V13700

The LM13600 and LM13700 have a subtle difference between their high-impedance output buffer biasing circuits that may give an edge to the LM13700 in high-performance audio applications. However, in every circuit I've built around the LM13600, I've been able to plug in the LM13700 with no changes to the circuit—thus, I consider them equivalent in my applications. Going forward, I will use the term LM13700 generically as a reference to any of the dual-transconductance op amps listed in **Table B-1***.*

LM13700 Chip Details

Let's discuss the purpose of each pin of transconductance op amp A in the package (Table B-2).

Table B-2. LM13700 Dual Transconductance Op Amp pin descriptions

Pin Number	Description
Pin 1	Amplifier Bias Input A
Pin 2	Diode Bias A
Pin 3	Noninverting Input A
Pin 4	Inverting Input A
Pin 5	Output A
Pin 6	Negative Supply
Pin 7	Buffer Input A
Pin 8	Buffer Output A
Pin 9	Buffer Output B
Pin 10	Buffer Input B
Pin 11	Positive Supply
Pin 12	Output B
Pin 13	Inverting Input B
Pin 14	Noninverting Input B
Pin 15	Diode Bias B
Pin 16	Amplifier Bias Input B

The two op amps are identical, and thus the discussion of op amp A applies to op amp B. You should download the National Semiconductor data sheet (November 1994) (*http://bit.ly/Y1xtMu*); it is *loaded* with all of the details and specifications of the chip, plus a collection of useful application circuits. If you want even more nitty-gritty, transistor-by-transistor details, search for scholarly papers related to the *LM13700* and *transconductance* in general.

Pin 1

Amplifier Bias Input A is the pin that sets the amplifier's actual *transconductance*, or the relation between the output and input currents. The current flowing into this pin controls how much current flows in and out of the chip's output. This is related to a factor applied to the differential voltage between the chip's noninverting and inverting inputs (pins 3 and 4, respectively). The maximum current recommended to apply to this pin is 2mA, and applying more than that could damage the chip.

Pin 2

Diode Bias A is where you apply current to bias the chip's linearizing diodes, which permit the chip to both accept a bit higher differential input voltage and improve its linearity and distortion figures significantly. Keeping the maximum differential input voltage between 10mV to 20mV allows you to leave this pin unconnected, because under those conditions, the chip provides far more than adequate distortion figures for synth-DIY work. Using the linearizing diodes improves the distortion figures dramatically, even with differential input voltage as high as 30mV. In my own applications I rarely use this pin and still find that the circuits deliver excellent performance.

Pin 3

Noninverting Input A is the noninverting input to the chip's front end differential amplifier. As mentioned above, when experimenting with this chip, it is important to keep the differential input voltage applied between this and pin 4 to between 10mV and 20mV for low distortion operation.

Pin 4

Inverting Input A is the inverting input to the chip's front end differential amplifier. As mentioned above, when experimenting with this chip, it is important to keep the differential input voltage applied between this and pin 4 to between 10mV and 20mV for low distortion operation.

Pin 5

Output A is the output of op amp A as a current, which is the result of the difference in input voltage between the inverting and noninverting inputs times the transconductance, which is controlled by the *Amplifier Bias Input A* on pin 1.

Pin 7

Buffer Input A is the base of the on-chip Darlington pair buffer. This input is relatively high impedance, but nowhere near as high as the gate of a JFET or noninverting input of a JFET input amplifier. For many applications the buffer is very useful, but in others you won't use it. Just leave this pin and pin 8 (*Buffer Output A*) unconnected in applications where you don't use the internal Darlington buffers.

Pin 8

Buffer Output A is the emitter output of the on-chip Darlington pair buffer. In practice, this output is typically tied to the negative supply through a resistor (4.7K to 20K or so), therefore the current supplied by the output is converted to a voltage. Just leave this pin and pin 7 (*Buffer Input A*) unconnected in applications where you don't use the internal Darlington buffers.

Three Examples of LM13700 Circuits

You'll find many interesting electronic music projects that use the LM13700 by visiting the Music From Outer Space website. I encourage you to use them as starting points for your own experimentation.

A Simple VCA Using the LM13700

In Figure B-2, we see half of an LM13700 used as a voltage-controlled amplifier. Notice that the input signal is capacitively coupled via C1 (a 1µF bipolar aluminum capacitor) and applied to the LM13700's noninverting input via the resistor attenuator R8 (an 82K resistor) and R9 (an 100 ohm resistor to ground). Pin 4 (the inverting input) is also tied to ground; thus if a 10V peak-to-peak signal (+/−5V oscillating about ground) is applied to the input, the differential voltage seen by the input is about 21mV peak to peak.

As discussed above, at this input level this VCA provides excellent operation with low distortion. In this application, we use the Darlington buffer; pin 5, *Output A*, connects to pin 7, *Buffer Input A*, which is also tied to ground via biasing resistor R10 (300K). Pin 8, *Buffer Output A*, is tied to V− via a 4.7K resistor (R11), converting the output current to a voltage that is capacitively coupled to the VCA's output via C2 (1µF bipolar aluminum capacitor).

U1-A is used as an inverting mixer (with a gain of one) for the control voltage inputs. U1-B, used as an inverting block with a gain of one, reinverts the mixed control voltage and drops it across the resistor attenuator comprised of R6 (47K resistor) and R7 (12K resistor to −9V). The output of U2-A (LM13700) is a product of the current flowing into pin 1, *Amplifier Bias Input A*, and the differential voltage at its inputs. Since pin 1, *Amplifier Bias Input A*, is referenced to V−, current only flows into it when the voltage presented to it is greater than V−. In this circuit, we simply bias *Amplifier Bias Input A* strongly toward V− with R7 (12K to −9V) and then use the control voltage (expected to be within +/–5V also) to raise the voltage, allowing current to flow through R6 (47K resistor), which increases the transconductance and as a result the amplitude of the voltage seen at the VCA's output. This is a great circuit to build on a solderless breadboard and experiment with.

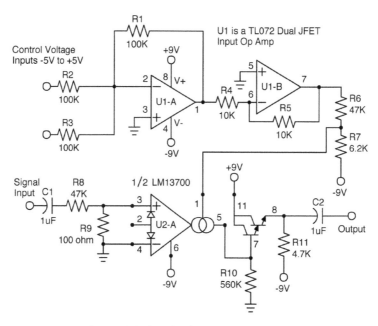

Figure B-2. *Simple VCA using the LM13700*

A Simple State-Variable VCF

This state-variable VCF is based on an application circuit presented in the *November 1994, National Semiconductor, LM13700/LM13700A Dual Operational Transconductance Amplifiers with Linearizing Diodes and Buffers* (*http://bit.ly/17fREqF*) data sheet. I started breadboarding using the component values suggested in the application note and then began experimenting. I followed the initial design but added a resonance control and adjusted the gains around the circuit to accommodate the voltage levels I wanted to use for control voltage, input signal, and output. The circuit presented in this section is found in the Music From Outer Space *Sound Lab Mini-Synth*, an analog synthesizer project many synth-DIYers enjoy building and using for experimental music.

Having two transconductance op amps in one chip is convenient for adding voltage control to a two-pole filter like the one presented in Figure B-3. The input level control helps keep the input signal in the optimum 10mV to 20mV peak-to-peak level. The transconductance op amps play the role of voltage-controlled integrators in this application. In order to provide voltage control, we employ the same biasing trick used in the VCA circuit presented previously.

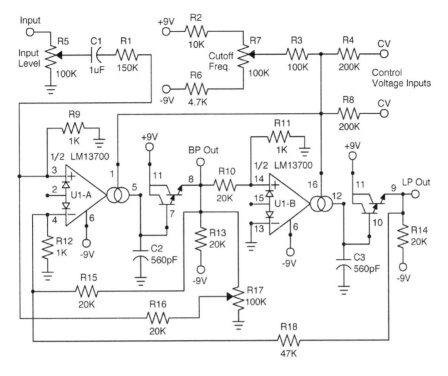

Figure B-3. *Simple state-variable VCF using the LM13700*

When the wiper of the Cutoff Frequency control is adjusted toward R6 (4.7K resistor to V–), the voltage on pins 1 and 16 (*Amplifier Bias Inputs*) is close to V–. Thus the current into pins 1 and 16 is minimal, resulting in low transconductance

through the op amps and low cutoff frequency. When the wiper of the Cutoff Frequency control is adjusted toward R2 (10K resistor to V+), the current into pins 1 and 16 goes up and the transconductance increases, resulting in higher cutoff

What's a Two-Pole Filter?

Voltage-controlled filters come in many flavors and designs. A VCF is classified by how many poles it has, or the number of decibels (dB) of attenuation it produces as the frequency of the signal passing through it increases by an octave. The more poles the filter has, the better able it is to pass only the harmonics in its passband and eliminate (or subtract) all others.

For example, popular filter specifications include "two pole," or 12dB per octave; and "four pole," or 24dB per octave. The low-pass VCF is one of the most prevalent found in analog synthesizers and passes frequencies that are below the cutoff frequency. The band-pass VCF passes a narrow range of frequencies whose center frequency is controlled with the cutoff frequency control. The high-

pass VCF passes frequencies that are above the cutoff frequency.

As a low-pass filter's cutoff frequency is swept from low to high, more and more high harmonics of the signal passing through it are heard. If the resonance is turned up, the characteristic "wah" sound is heard due to the emphasized sinusoidal cutoff frequency adding to the signal. As a high-pass filter's cutoff frequency is swept from low to high, less and less low harmonics of the signal passing through it are heard. As you have probably guessed, as the band-pass filter's cutoff frequency is swept from low to high, the band of harmonics being passed becomes higher and higher.

frequency. In this design, the cutoff frequency is directly related to the control voltage. Additional modulation of the cutoff frequency is possible by applying voltage to either or both control voltage inputs. As control voltage applied to either or both of these inputs increases, current via resistors R4 and/or R8 into pins 1 and 16 goes up, resulting in increased transconductance and higher cutoff frequency.

The control voltage (CV) inputs can be modulated with up to 5V peak to peak. The filter expects an input signal with an amplitude of about 3V to 5V peak to peak. One of the coolest things about this VCF is that it provides both low-pass (pin 9 of U1-B) and band-pass (pin 8 of U1-B) response outputs. I encourage you to experiment with this circuit on your solderless breadboard, change values and see what happens. For instance, changing the value of the integrator caps C2 and C3 (but keeping their values equal) will result in changing the overall range of the filter's cutoff frequency. Higher value caps (.001µF, for example) will lower the overall range, and lower value caps (330pF, for example) will raise the entire range.

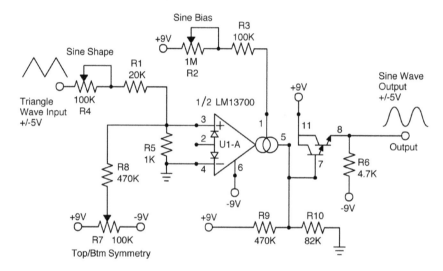

Figure B-4. *Triangle-to-sine converter using the LM13700*

A Triangle-to-Sine Converter

This last example, shown in Figure B-4, demonstrates using the LM13700 to convert a triangle wave into a sine wave. A useful side effect of overdriving the input of the LM13700 with a triangle wave is that the resulting nonlinear distortion looks amazingly like a sine wave. In this circuit, the input is expected to be a triangle wave oscillating about ground with amplitude of +/–5V.

The triangle wave is applied to the noninverting input (pin 3) of U1-A (1/2 LM13700) via trim pot R4 (100K) and resistor R1 (20K), where it is drop-ped across R5 (1K resistor to ground). The inverting input of U1-A (pin 4) is connected directly to ground. Even with R4 adjusted to place as much resistance as possible between the input and R1, a 10V peak-to-peak signal (+/–5V) will result in about +/–41mV between U1-A's differential inputs. At this amplitude, we are already driving the output into nonlinear distortion territory.

Adjust R4 (Sine Shape) until the distorted triangle on pin 8 looks sinusoidal. Trimmer R2 (Sine Bias) affects both the nonlinear distortion and the amplitude of the output signal. The adjustments of trimmers R1 and R2 are somewhat interactive, so you'll need to play with their settings while observing U1-A pin 8 on an oscilloscope until you obtain the best sinusoidal approximation. Trimmer R7 (Top/Btm Symmetry) adjusts the top-to-bottom symmetry of the wave's shape and allows you to center the output about ground. Adjust the three trimmers until the lowest distortion sine wave is observed at the output.

I can't emphasize enough the value of getting *very* familiar with this chip. Search out the National Semiconductor application notes on the LM13700 I suggested above, study the circuits, and experiment with those that interest you using your solderless breadboard.

Working with CMOS Logic Chips

There are many families of logic chips around, but CMOS takes the prize for low power operation, high noise immunity, wide power supply operation, and wide range of available functions. It is important to note that the CMOS family I'm referring to is the CD4XXXX family. There are many equivalent chips in the 74CXXXX family, but always check the power supply specifications to make sure that the chip you're interested in works at the supply voltage you're using. Some 74CXXXX and all 74HCXXXX (high-speed CMOS) family chips have a far more limited supply range (2V to 6V) and if exposed to higher voltages will act erratically, get hot, and more than likely be destroyed. Also, just so you know, CMOS chips use the power pin nomenclature VDD (also VCC) for V+ and VSS (also ground) for ground.

It's important to observe static precautions when working with CMOS, as it is far more susceptible to static damage than hardier bipolar transistor-based logic families like TTL or LS. CMOS uses metal-oxide semiconductors (MOS) which, due to their physical construction, are more prone to serious static damage with mishandling. To avoid problems, make sure your workstation isn't functioning as a Van de Graaff generator. If you notice static snaps between you and items in your work area, you need to get a static mat and strap installed before you start working with CMOS. Always store your CMOS components in protective static foam and protective static bags, and make sure you're grounded via a static strap and mat when handling them.

CMOS gate inputs have very high input impedance (10^{12} ohms). Therefore, when experimenting or designing a circuit using CMOS you *must* connect any unused inputs to either VDD or VSS. If you don't, they can be influenced by electrical fields or tiny leakage currents and change states when you least expect it or oscillate, drawing significant current. It is important to tie unused CMOS inputs to the *appropriate* logic level (VDD or VSS) such that it does not interfere with the

chip's logical function. For example, if you have a four-input OR gate and you only use three of the gate's inputs in your logic, tie the fourth input to ground, since tying it to VDD will cause the OR gate's output to ignore the other inputs, and always output a high logic level. As another example, you wouldn't want to tie a counter's unused reset pin to the logic level that inhibits it from counting.

The logical output levels of a CMOS gate are basically equal to the plus and ground power rails. For example, if your circuit is powered with +12V and ground, the output of the gates will be very close to +12V when high and very close to ground when low. CMOS gates are not designed to source or sink large currents, and the values for these parameters on most data sheets are usually in the 0.3mA to 2mA range, depending on supply voltage. Those are the values of current that a CMOS gate's output can source or sink even under maximum fanout and maximum speed.

In the real world, when all you want to do is drive an LED, a CMOS output can easily source 5mA to 10mA, but the high and/or low logic voltage levels will be affected due to voltage dropped across the gate's MOS output transistors. For example, when using a CD40106 (powered with +12V and ground) to drive a general purpose LED with a 1K current-limiting resistor, the gate-sourced 8.75mA and its high logic voltage level dropped to 10.2V. When I changed the setup so that the output of the gate was sinking the current, I observed 9mA flowing through the LED, but the gate's low logic voltage level rose to 1.23V. Therefore, when using a CMOS gate's output to drive an LED via a current-limiting resistor, I recommend that you *don't* use the same gate's output to drive other logic.

Logic Chips in Synth-DIY

You can do a lot of cool tricks with CMOS gates that aren't possible with other logic families because of their high input impedance, high noise immunity, current source and sink capability, and VDD and ground level outputs. I'm going to present some of the CMOS ICs I use often in my analog synth designs along with some comments and tips for using them. I'm including pinout diagrams for each so you'll have them right here when you need them. I'm saving the versatile CD40106 hex inverter Schmitt trigger for last, and I'll show you how to get a lot of mileage out of it in your designs.

CD4013 Dual D Flip-Flop

The CD4013 contains two D type flip-flops. D flip-flops have many uses for synth-DIY including frequency division, switch debouncing, data latching, and more. You definitely want some of these gems in your parts cabinet.

Each flip-flop has Q and NOT-Q outputs, Set and Reset pins that can be used to force the flip-flop into one or the other states, and Data and Clock inputs used for latching and frequency division applications. The CD4013's pinout is shown in Figure C-1.

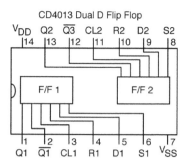

Figure C-1. *CD4013 dual D flip-flop IC diagram*

The circuit in Figure C-2 demonstrates two modes of use for the CD4013. U1-A is used as a push-button debouncer, and U1-B is used as a divide-by-two counter. In the push-button debouncer, we use U1-A's Set and Reset (pins 6 and 4, respectively) inputs, and we ground the Data and Clock inputs (pins 5 and 3 respectively) since they're not being used. Not grounding unused pins 5 and 3 would leave them susceptible to leakage currents or electrical fields due to their 10^{12} ohm input impedance. Unconnected CMOS inputs will cause odd, hard-to-track-down problems with your circuits, so never leave them floating. Upon pulsing U1-A's Set input (pin 6), the Q output of U1-A (pin 1) goes high and charges C3 (.1μF ceramic capacitor) via R4 (300K resistor) until the voltage on U1-A's Reset input (pin 4) is high enough to reset it, causing U1-A's Q output (pin 1) to return low and its NOT-Q output (pin 2) to go high.

U1-A's Set input (pin 6) is held at ground by R2 (1M resistor). In order to pulse this input and *set* the flip-flop (make the Q output go high), we use push-button S1. One terminal of S1 is connected to +V and the other is connected to R1 (20K resistor). When the push button is pressed, current flows from +V through R1 and charges C1 (.001μF ceramic cap) very quickly. C1 and R1 help filter the switch bounce as S1 is pressed. The front end of the quickly rising voltage on C1 passes through C2 and is dropped on R2 (1M resistor), creating a positive pulse that sets flip-flop U1-A. In the set state, U1-A's Q output (pin 1) is high and its NOT-Q (pin 2) output is low. U1-A's high Q output (pin 1) charges C3 via R4 until the voltage on U1-A's Reset input (pin 4) is high enough to reset the flip-flop. In the reset state, U1-A's Q output (pin 1) is low and its NOT-Q (pin 2) output is high. As a result, every time the push button (S1) is pressed, a short positive digital pulse occurs on the Q output of U1-A. Push-button contacts *bounce* when released just as much as they do when pressed. C1 (.001µF ceramic capacitor), which charges to +V via R1 on push-button press, filters the switch bounce noise, preventing spurious pulses on push-button release.

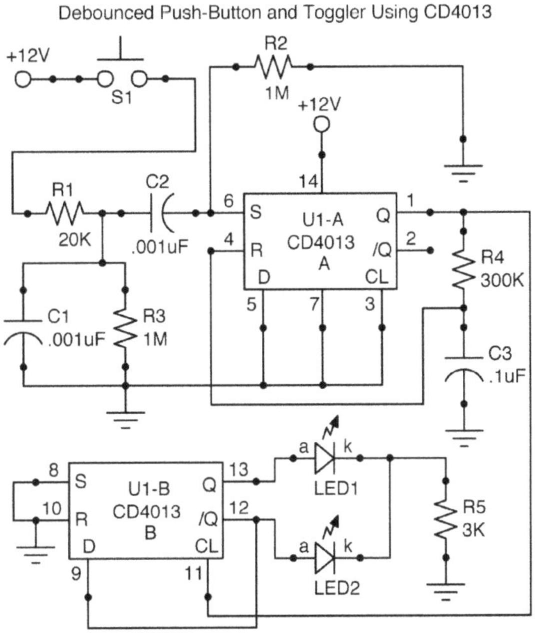

Figure C-2. *Debounced push button and toggler using CD4013*

U1-B is used as a divide-by-two counter. In this circuit, we ground U1-B's Set and Reset inputs (pins 8 and 10, respectively) and use its Data and Clock inputs (pins 9 and 11, respectively). Notice that U1-B's NOT-Q output (pin 12) is connected to its Data input (pin 9). The Data and Clock inputs are what give the CD4013 its ability to be used as a data latch. When the Data input is held at logic 1, for example, and the Clock input is clocked with a positive-going pulse, the Q output will assume the logic level that was present

on the D input, logic 1 in this case. If the Data input had been held low during the clock, then a logic 0 would be latched to the Q output. Anytime a flip-flop's Q output is high, its NOT-Q output is low, and vice versa.

Since U1-B's NOT-Q output (pin 12) is connected to its Data input (pin 9), U1-B's Q output (pin 13) will assume the logic level on the Data input when the signal applied to its Clock input (pin 11) goes high. Immediately afterward, U1-B's NOT-Q output (pin 12) goes to the logic level opposite that of the Q output. Consequently, if U1-B's NOT-Q output was low during the clock's low-to-high transition, then its Q output will assume a low logic level, and NOT-Q will immediately go to a high logic level. What will happen if we clock U1-B's Clock input now? U1-B's Q output will go high, and its NOT-Q output will immediately go low. Since it takes one low-to-high clock transition to bring U1-B's Q high and one low-to-high clock transition to bring it low again, we are effectively dividing the incoming clock frequency by 2.

If we applied a 1 kHz clock to the clock input of a flip-flop wired in the same manner as U1-B, we would see a 500 Hz square wave at the Q output (clock divided by 2). In the case of this circuit, we're just using the debounced switch and resulting pulse to toggle the state of U1-B. One press brings U1-B's Q output high (and NOT-Q low), and the next press brings U1-B's Q output low (and NOT-Q high). For each press of S1, the LEDs on the Q and NOT-Q outputs will change states.

If you needed more than two states, you could use the debounced switch portion of the circuit (U1-A and associated components) to pulse the clock input of a CD4017, which would give you up to 10 push-button selectable states. Those high outputs could be controlling CD4066 analog switches to route signals or whatever you need.

CD4011 Quad Two-Input NAND Gate

The quad two-input NAND gate, whose schematic symbol is shown in Figure C-3, is a staple of any CMOS parts cabinet. Table C-1 shows the truth table for the two-input NAND gate. You can easily get the AND function by simply connecting the output of one gate to both inputs of a second gate and then using the output of the second gate. Before you bring another IC into your design, remember that the CD40106 used with some high-speed diodes and a resistor can emulate several logic functions. There's a lot more about this in the section "CD40106 Hex Inverting Schmitt Trigger " on page 156.

Table C-1. NAND gate truth table

Input 1	Input 2	Output
0	0	1
0	1	1
1	0	1
1	1	0

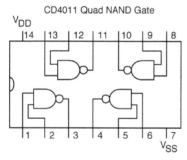

Figure C-3. *CD4011 quad two-input NAND gate IC diagram*

CD4001 Quad Two-Input NOR Gate

The quad two-input NOR gate, whose schematic symbol is shown in Figure C-4, is another staple of any CMOS parts cabinet. Table C-2 shows the truth table for the two-input NOR gate. You can easily get the OR function by simply connecting the output of one gate to both inputs of a second gate and then using the output of the second

gate. Before you bring another IC into your design, remember that the CD40106 used with some high speed diodes and a resistor can emulate several logic functions. There's a lot more about this in the section "CD40106 Hex Inverting Schmitt Trigger " on page 156.

Table C-2. NOR gate truth table

Input 1	Input 2	Output
0	0	1
0	1	0
1	0	0
1	1	0

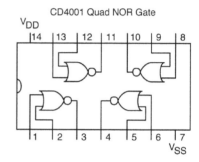

Figure C-4. *CD4001 quad two-input NOR gate IC diagram*

CD4066 Quad Analog Switch

The CMOS CD4066 quad analog switch (Figure C-5) is a package of four electronic switches. Each switch can pass analog signals between its two I/O pins whenever its control pin is at logic 1. These switches work really well in sample and hold circuits and electronic signal routers. It is important to remember that the CD4066 is a single-supply device that expects the voltage levels passing through it to be in the ground to +V range. You can't put signals through the device with voltage valleys lower than the VSS pin's voltage, or parasitic and/or protection diodes on the chip will turn on, distorting and/or clipping the signal. If you want to use one of the analog switches to switch an op amp output, you must bias the op amp's output voltage up so that the signal is operating about 1/2 +V (instead of about ground). It is also important to ensure that the op amp's amplitude will not exceed ground to +V levels. See "Biasing the Op Amp's Output" on page 128 in Appendix A for information related to setting op amp gain and bias levels. These analog switches have a typical on resistance ranging between 270 ohms (5V operation) and 80 ohms (15V operation). In the off state, the switches are specified to have leakage current in the picoamps, meaning they are really *off*.

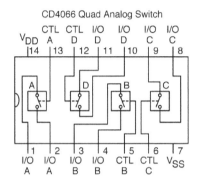

Figure C-5. *CD4066 quad analog switch IC diagram*

CD4024 Seven-Stage Ripple Carry Binary Counter

This is a general purpose binary up counter with a clock input, reset line, and seven outputs. The CD4024's schematic symbol is shown in Figure C-6. The binary outputs count up on the negative edge of the clock. Hitting the reset line with a logic 1 will reset the counter to 0. This chip is great for when you want to generate several square wave suboctaves from an oscillator. It's also great to use with an R/2R ladder to generate a nice, stepped ramp wave. By switching the outputs to the R/2R ladder, you can change the increments by which the voltage changes as the count advances.

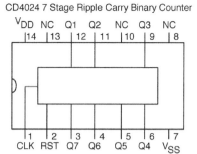

Figure C-6. *CD4024 seven-stage ripple carry binary counter IC diagram*

CD4042 Quad D Latch

The quad D latch, shown in Figure C-7, is what you use when you need to hold on to a 4-bit value. The 4 bits of data are presented to the chip's four data input lines (D1 thru D4). If the Polarity pin (POL) is held at logic 0, the data on the Data inputs is transferred to the Q and NOT-Q outputs when the Clock input (CL) is at a low level. When the Clock input returns high, the data is latched into the Q and NOT-Q outputs until the clock goes low again. If the Polarity input is held at logic 1, then the data is transferred from D to Q and NOT-Q when the Clock input is high. When the Clock input returns low, the data is latched into the Q and NOT-Q outputs until the clock goes high again. As expected, if a logic 1 is present on a Data line when the clock permits data to be transferred and latched, the Q output for that latch will be at logic 1, and its companion NOT-Q will be at logic 0.

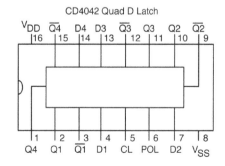

Figure C-7. *CD4042 quad D latch IC diagram*

CD4094 Eight-Stage Shift Register

The CD4094 eight-stage shift register, shown in Figure C-8, contains an eight-stage shift register whose outputs are fed to an 8-bit latch. Logic data presented to the chip's Data input (DATA) is serially shifted through the shift register on the positive edges of the chip's Clock input (CL). The states of the internal shift registers can be latched to the chip's internal 8-bit latch by applying a negative edge to the Strobe input (STB). The eight internal latched outputs feed eight internal tri-state buffers that present the data to their outputs (Q1 thru Q8) whenever the Output Enable input (EN) is at logic 1. A three-state output can be logic 1, logic 0, or a special third state, *floating*, which acts as if the output has been disconnected.

You can have a lot of analog synth fun with one of these by connecting an R/2R ladder to the output while clocking in a string of random ones and zeros. Each time the Strobe pin sees a negative edge, the random data flowing through the internal shift register gets latched into the internal 8-bit latch and presented to the R/2R ladder. Using an 8-bit R/2R ladder will provide the possibility of 256 discrete voltage levels. The voltage at the output of the R/2R ladder can be used to drive oscillators, filters, or any voltage-controlled module. You could also set up a long-cycle pseudo white noise generator by using some XOR gates and a few of these chips, since they're designed to be serially cascaded with additional CD4094 chips.

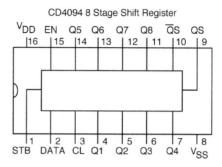

Figure C-8. *CD4094 eight-stage shift register IC diagram*

CD40193 Binary Up/Down Counter

The CD40193 binary up/down counter, shown in Figure C-9, contains one 4-bit binary counter that can count up, count down, be preloaded with a count, or be reset. This counter can easily be cascaded with additional CD40193s by connecting the Carry (CY) and Borrow (BW) outputs to the next CD40193's Count Up and Count Down inputs, respectively. The counter counts up or down depending on which clock input is used, while the other one is held at a high logic level. If the Count Up input (CL UP) sees a positive-going clock edge while the Count Down input (CL DN) is held high, the counter will count up. If the Count Down input (CL DN) sees a positive-going clock edge while the Count Up input (CL UP) is held high, the counter will count down. In order to *jam* a count into the counter and make it count up or down from there, you present the 4-bit count to the counter's four data lines (DA thru DD) and then bring the Load input (LD) momentarily low. The Clear input (CLR) forces the counter's outputs (QA thru QD) to 0 when brought to a logical one.

This counter is great for making sequencers that use the CD4514 and/or CD4067 chips, since both chips require 4 bits of data to make one of their 16 outputs active. This counter is used in the MFOS 16-Step Sequencer, MFOS Voltage Quantizer, and MFOS Vari-Clock Module, to name just a few. Check out the projects to see the chip in action.

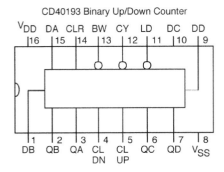

Figure C-9. *CD40193 binary up/down counter IC diagram*

CD4514 4-Bit Latch/4-to-16 Line Decoder

The 4-bit latch/4-to-16 line decoder (see Figure C-10) has four inputs (D1 thru D4), which are used to select which of the chip's 16 outputs (S0 thru S15) will be at logic 1. All nonselected outputs go to a low logic level. The chip has an internal 4-bit latch to hold the 4-bit selection, if necessary. The chip's Strobe input (STB) controls whether or not the data is latched. When the strobe input is high, the 4 bits on the select lines pass through to the internal multiplexer. When the strobe input goes low, however, the last 4-bit code presented while strobe was high is held in the internal latch, keeping the selected output high. If the chip's Inhibit input (INH) is brought to logic 1, all of the outputs (S0 thru S15) go low. In normal operation, Inhibit should be held low. If Strobe is held at logic 1, the state of the outputs (S0 through S15) will correspond to the data presented to select inputs (D1 thru D4).

This chip is used in the Music From Outer Space 16-Step Sequencer to drive the channel indicator LEDs and to generate the selectable gate outputs. It is controlled by the previously presented CD40193 binary up/down Counter. I hope you can see the potential for putting this chip to work in your own synth-DIY work.

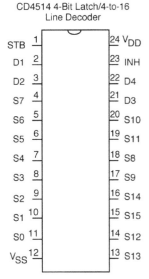

Figure C-10. *CD4514 4-bit latch/4-to-16 line decoder IC diagram*

CD4017 Decade Counter with 10 Decoded Outputs

The CD4017 decade counter (Figure C-11) with 10 decoded outputs has 10 outputs (0 thru 9) that go high sequentially every time the chip's Clock input sees a low-to-high clock transition while the Clock Inhibit input (CIN) is held low. If the Clock Inhibit input (CIN) is brought to logic 1, the Clock (CLK) input is ignored. The chip has a Reset input (RST), which when brought to logic 1 causes the 0 output to go high and all others to go low. Only one output at a time is at logic 1 while all others are at logic 0.

This chip is used in the simpler Music From Outer Space 10-Step Sequencer to drive the channel indicator LEDs, generate the selectable gate outputs, and bias the channel voltage set pots. For making a simple synth-DIY sequencer, you won't find a more appropriate chip.

CD4017 Decade Counter with 10 Decoded Outputs

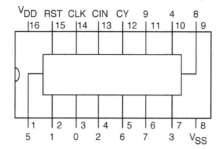

Figure C-11. *CD4017 decade counter with 10 decoded outputs*

CD4067 Single 16-Channel Multiplexer/Demuliplexer

The CD4067 single 16-channel multiplexer/demultiplexer, shown in Figure C-12, contains 16 analog switches, all of which close to a common point (CMN I/O). Any one of the chip's 16 I/O ports (I/O 0 thru I/O 15) can be selected via the chip's four select lines (A thru D). If a logic 1 is applied to the chip's Inhibit input (INH), none of the 16 I/O ports is connected to the common output. The on resistance for these chips is around 470 ohms when powered by 5V and 125 ohms when powered by 15V. The off resistance is specified as leakage current and is in the nanoamps.

This chip is used in the Music From Outer Space 16-Step Sequencer to route the voltage from the coarse and fine tuning pots to the output of the sequencer. This is another great chip for DIY experimentation related to sequencers.

CD4067 Single 16 Channel Multiplexer/Demultiplexer

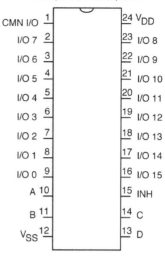

Figure C-12. *CD4067 single 16-channel multiplexer/demultiplexer IC diagram*

CD40106 Hex Inverting Schmitt Trigger

The CD40106 hex inverting Schmitt trigger contains six inverters (see Figure C-13). The thing that makes this chip special is that the inverters are endowed with hysteresis. Once the input has gone low enough to cause the output to go high, you have to bring the input voltage to above 66% of the VDD voltage to get the output to go low again. Once the input voltage has gone high enough to bring the output low, you need to bring the input voltage to less than 33% of VDD to get the output to go high again. The hysteresis zone is typically 1/3 of VDD. Hysteresis makes these inverters as immune to circuit noise as you can get, but also allows them to be used as impromptu gates of amazing diversity.

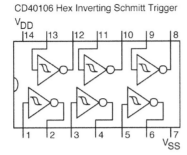

Figure C-13. *CD40106 hex inverting Schmitt trigger IC diagram*

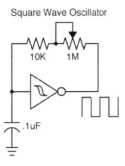

Figure C-14. *CD40106 square wave oscillator*

Let's look at some of the cool things you can do with this chip. Remember that these circuits expect CMOS outputs to be fed to their inputs. They all rely on CMOS's VDD and VSS level outputs and their ability to both source and sink current to work correctly.

CD40106 Square Wave Oscillator

Figure C-14 presents the schematic for a simple square wave oscillator made with one CD40106 Schmitt trigger inverter. The output is fed back to the input via a 10K resistor and a 1M potentiometer wired as a variable resistor. The input has a capacitor hanging off of it, and since this is an inverter, we know that when the input is high, the output is low, and vice versa. If on power-up the output is low, it will pull current through the resistors in the feedback network and eventually drag the input below the schmitt inverter's low threshold voltage, at which time the output will snap high. With the output high, current will flow through the resistors in the feedback network and charge the cap on the input until it goes above the Schmitt inverter's high threshold voltage, at which time the output will snap low again. This cycle repeats, resulting in square wave oscillation at the output of the inverter. A rough estimate of the frequency when using this configuration is 0.75 x 1/(RC). The potentiometer in the feedback network can be used to vary the frequency over a fairly wide range. Experiment on your solderless breadboard with different values of resistors and capacitors to see how the frequency changes.

CD40106 Pulse Wave Oscillator

The pulse wave oscillator, shown in Figure C-15, is practically the same as the square wave oscillator, except for the diode in the feedback network. When the output of the inverter is high, current rushes through the forward-biased 1N914 diode and rapidly charges the capacitor on the input. When the output is low, the diode is reverse-biased and the current flows through the 10K resistor and 1M pot wired as a variable resistor in the feedback path. Thus, the high time will be very short, and the low time will depend on the setting of the 1M potentiometer in the feedback network. Listen to the difference between the square wave and the pulse wave. Try different values of capacitors and resistors to see the effect on the oscillator's frequency. Can you guess what happens if you turn the diode around? Get out your solderless breadboard and find out.

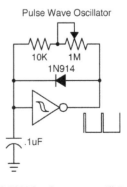

Figure C-15. *CD40106 pulse wave oscillator*

CD40106 Multiple-Input NAND Gate

If you ever need a NAND gate with as many inputs as you want, you can use a leftover CD40106 inverter instead of adding a CD4011 chip to your design (see Figure C-16). And, *surprise*, putting another inverter on the output turns this into a multiple-input AND gate.

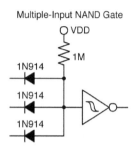

Figure C-16. *CD40106 multiple-input NAND gate*

CD40106 Multiple-Input NOR Gate

If you ever need a NOR gate with as many inputs as you want, you can use a leftover CD40106 inverter instead of adding a CD4001 chip to your design (see Figure C-17). And, *surprise again*, putting another inverter on the output turns this into a multiple-input OR gate.

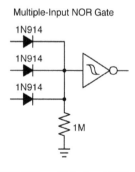

Figure C-17. *CD40106 multiple-input NOR gate*

CD40106 Positive Edge to Negative-Going Pulse

When working with logic circuits, you often need to generate a narrow negative (or positive pulse) in response to the rising edge of a logic signal. The circuit presented in Figure C-18 is perfect for

that. With no changes occurring on the input, the 1M resistor to ground holds the input low, and thus the output high. When a low-to-high logic transition is applied to the input, the front of the voltage step is capacitively coupled through the .001µF cap and dropped onto the 1M resistor as a narrow low-to-high pulse. The output of the Schmitt inverter pulses low in response and then returns high due to the 1M resistor holding the input at ground. Adding a Schmitt inverter to the output would generate a positive pulse in response to the low-to-high transition on the input.

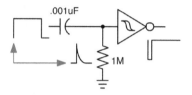

Figure C-18. *CD40106 positive edge to negative-going pulse*

CD40106 Negative Edge to Positive-Going Pulse

Likewise, when working with logic circuits, you just as often need to generate a narrow positive (or negative pulse) in response to the falling edge of a logic signal. The circuit in Figure C-19 is perfect for that. With no changes occurring on the input, the 1M resistor to VDD holds the input high and thus the output low. When a high-to-low logic transition is applied to the input, the front of the voltage step is capacitively coupled through the .001µF cap and dropped onto the 1M resistor as a narrow high-to-low pulse. The output of the Schmitt inverter pulses high in response and then returns low due to the 1M resistor holding the input at VDD. Adding a Schmitt inverter to the output would generate a negative pulse in response to the high-to-low transition on the input.

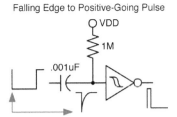

Falling Edge to Positive-Going Pulse

Figure C-19. *CD40106 negative edge to positive-going pulse*

CD40106 Narrow Pulse Stretcher

The circuit in Figure C-20 is perfect for times when you have a pulse that's too narrow for your needs. Maybe you want to flash an LED in response to the pulse, but the pulse itself is no good because the LED goes on and off so fast that it's hardly visible. You need a pulse stretcher. The narrow positive-going pulse on the input is inverted to become a narrow low-going pulse. During the low-going pulse, the 1N914 diode is forward biased and rapidly discharges the capacitor on the second Schmitt inverter's input. When the cap discharges, the output of the second inverter goes high and stays there until current through the 1M resistor recharges the cap, making the output go low again. By making the resistor higher in value, you will make the pulse longer, and vice versa. You can also make the cap

larger to increase the length of the stretched pulse, but at some point, if the input pulse is really narrow, you won't have enough time to discharge it through the 1N914.

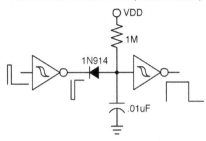

Narrow Pulse to Wider Pulse (Pulse Stretcher)

Figure C-20. *CD40106 narrow pulse stretcher*

I hope that this brief introduction to CMOS gives you the courage to get some chips and experiment with them on your solderless breadboard. In your work with synth-DIY, you'll run into CMOS again and again, and now you'll know more about it. If you can lay your hands on either a real National Semiconductor CMOS Data Book or an electronic copy on CD or DVD, by all means do so. Reading through the wealth of information in the IC data sheets and application notes will give you a leg up in your future design work.

Index

Symbols

9V battery eliminators, 95
9V battery snap polarity, 77

A

AC mixers, 53
AC vs. DC coupling, 53
AC-coupled audio mixers, 131
ADSREG modules (see attack decay sustain release (ADSR) envelope generator modules)
Alien Screamer synthesizer (Magnaval), 3
All Electronics, 62
aluminum chassis, 61
aluminum electrolytic capacitors, 24
amplitude envelopes, 48
amplitude modulation, 55
analog multiplier chips, 55
analog oscilloscopes, 9–11
analog switch (CD4066), 129
analog synthesizer-DIY, 3, 150
analog synthesizers
    categories of, 37

configuration of, 57
defined, 2
voltage control of, 38
vs. digital synthesizers, 1
analog, defined, 2
AR Mod Depth control, 59
AREG module (see attack release (AR) envelope generator modules)
arpeggio, 57
Art of Electronic Sound project, 118
Attack control, 60
attack decay sustain release (ADSR) envelope generator modules, 45, 48, 49–51
attack release (AR) envelope generator modules, 48, 49, 81, 96
Attack Time control, 49
Audacity
    edit tools, 114
    effect tails, 116
    effects, 114
    exporting projects, 112
    file formats, 112
    interface, 111
    introduction, 111
    Mixer Board, 121

Mixer Board window, 112
Multi-Tool Mode, 113
saving projects, 112
sound card discovery, 111
sound-on-sound recording, 114
tools/toolbar, 113
transport controls, 113
audio amplifiers, 105
audio interfaces, 107
audio signal mixers, 54, 108, 131

B

banana jacks, 37
bandwidth, 10
bench power supplies, 14
bias/biasing, 129
BIFET input op amps, 93
bill of materials (BOM)
    audio interfaces, 109
    Noise Toaster, 62
binarily weighted resistor DAC, 136
binary counter (CD4024), 136
binary up/down counter (CD40193), 154
bipolar electrolytic caps, 24

F

feedback resistors (RFBs), 87, 90
film capacitors, 24, 25
4-bit latch/4-to-16 line decoder (CD4514), 155
frequency counters, 17
frequency response, 10
frequency sweep capability, 18
full-wave rectified dual-analog power supplies, 15
function generators, 18

G

gain formulae, 88, 89, 127, 132
gain, setting for buffers, 128
gate truth tables, 152
gates, 39, 49, 50
glide controls, 47
goggles, 32

H

hand tools, 31
hard sync, 43
helping hands alligator clips device, 67
Hertz (Hz), 42
hex Schmitt inverter (CD40106), 139, 156
high impedance, 86
high-impedance buffer follower, 128
high-level signals, 32
hobby knives, 31
hybrid synthesizers, 37
hysteresis (positive feedback), 91–93
hysteresis zone, 92, 93

I

input channels, 10
input resistor (RIN), 87
inputs, 110
integral stereo panning mixers, 53
integrated circuit (IC) removers, 31
integrated circuit (IC) sockets, 67
integrated circuits (ICs), 29, 62, 67
inverting buffer gain formula, 88
inverting buffers, 86, 127

J

Jameco Electronics, 62
JFET transistors, 29
joining clips (Audacity), 120
joystick controls, 47

K

keyboard controllers, 39, 47
keyboard voltage modulation, 44
knobs, 27

L

leads, soldering, 22, 67
LED ring clips, 72
LEDs, 72, 130, 137–140
legato, 50
legend overlays, 70
LFO Mod Depth control, 59
LFO modules, 38, 46, 81, 97
line input, 107
linear control voltage inputs, 42
LM13700 Dual Transconductance Op Amp
    chip details/pin descriptions, 142
    manufacturers, 141
    schematic, 141
    triangle-to-sine wave converter, 146
    used as state-variable VCF, 144
    used as VAC, 143
logic chips, 34, 149–159
logic probes, 13
low note priority, 47
Low-Frequency Oscillator (LFO) modules (see LFO modules)

M

Magnaval, Bernard, 3, 5
Make: Electronics (Pratt), xiii
Manual Gate button, 60

MFOS (Music From Outer Space), xi
MFOS 16-Step Sequencer, 57, 154, 155, 156
MFOS Noise Toaster (see Noise Toaster)
MFOS Sound Lab Mini-Synth, x, 144
MFOS Vari-Clock module, 154
MFOS Voltage Quantizer, 135
MFOS Web SFZ Helper application, 125
MFOS website, xi, 29
Mixer Board (Audacity), 121
mixers, 53
Mod Depth control, 59
Mod Source switch, 60
modular analog synthesizers, 6, 37, 57
modulation inputs, 42
modulation wheels, 47
mono keyboard controllers, 47
Moog, Robert, 1
Mouser Electronics, 62
Multi-Tool Mode (Audacity), 113
multichannel DC modulation mixer, 132
multichannel oscilloscopes, 10
multichannel signal mixers, 44
multimeters, 12, 16, 77
multiple-input NAND gate (CD40106), 158
multiple-input NOR gate (CD40106), 158
Multitrack Layering exercise, 107, 121
Multitrack Volume Envelope Layering exercise, 107, 122–125
Music From Outer Space (MFOS), xi, 29
Musique Concrete exercise, 107, 118–121
Mute control (Audacity), 114

N

N-Channel JFET transistors, 29
narrow pulse stretcher (CD40106), 159
needle-nose pliers, 31

state variable VCFs, 43
static precautions, 31, 32, 149
subtractive synthesis elements, 43
surplus electronic parts, 24
switches, 27, 62
synchronized VCOs, 43
synth-DIY, ix, 3, 150
synth-DIY tools (see tools)
synthesizers, defined, 3

tantalum electrolytic capacitors, 24, 32
test leads, 31
Thurm, Michael, 6
Time Shift Tool (Audacity), 114, 119
tools, 9
    (see also electronic components)
    bench power supplies, 14
    capacitance meters, 16
    digital multimeters, 12, 16
    frequency counters, 17
    function generators, 18
    hand, 31
    logic probes, 13
    oscilloscopes, 9–11
Track menu (Audacity), 115
transconductance op amp, 141
transformers, 28
transistors, 29, 138
triangle (tri) core oscillators, 43
triangle wave, 41, 52

triggers, 39, 49, 50
Trim tool (Audacity), 114
trimpot adjustment tool, 31
troubleshooting, 31–35, 82
tweezers, 31
two-pole filters, 145

Vari-Clock module, 154
VCA modules, 38, 45, 80, 104
VCF Cutoff Frequency control, 59
VCF modules, 38, 39, 43–45, 79, 103
VCO frequency doubling, 38
VCO frequency modulation, 42
VCO modules, 38, 39, 40–43, 78, 99–102
vertical sensitivity, 10
vias, 65
vibrato, 42, 46
virtual ground, 14
Vocoder effect (Audacity), 115
voltage control of VCO frequency, 42, 43
voltage mixer and exponential converter, 99
Voltage Quantizer, 135
voltage sequencers, 56
Voltage-Controlled Amplifier (VCA) modules, 38, 45, 80, 104
Voltage-Controlled Filter (VCF) modules, 38, 39, 43–45, 79, 103

Voltage-Controlled Oscillator (VCO) modules, 38, 39, 40–43, 78, 99–102

W

wall warts, 95
waveforms, 41, 41, 97
Web SFZ Helper application, 125
Weird Sound Generator, 4
white noise, 53
white noise generator modules, 51, 102
white noise waveform, 51
window comparators, 130
Winter (Carlos), 121
wire strippers, 31
wiring diagrams
    Noise Toaster, 71, 73
    sound card audio interface, 109, 110
workbench tools (see tools)
working voltage, 25

X

XOR logic gates, 51

Z

Z wire, 65
Zoom Tool (Audacity), 113, 114

About the Author

Ray has been interested in analog synthesizers since the first time he heard *Switched-On Bach* back in 1968. That magic box on the cover of the album with all of the knobs, switches, and patch cords grabbed his attention and never let it go. After working at U.S. Steel, Intec Systems, Siemens Pacesetter, and Telectronics, he now runs his popular website, Music From Outer Space, full-time. Most of his electronics learning has been hard-won and experiential, with hundreds of hours devoted to reading, breadboarding, experimenting, and appreciating analog synthesis.

Colophon

The cover and body font is BentonSans, the heading font is Serifa, and the code font is Bitstreams Vera Sans Mono.

CPSIA information can be obtained
at www.ICGtesting.com
Printed in the USA
BVHW011959261022
650375BV00001B/9

9 781449 345228